SpringerBriefs in Molecular Science

More information about this series at http://www.springer.com/series/8898

Abhijit Bandyopadhyay · Tamalika Das
Sabina Yeasmin

Nanoparticles in Lung Cancer Therapy - Recent Trends

 Springer

Abhijit Bandyopadhyay
Tamalika Das
Sabina Yeasmin
Department of Polymer Science
  and Technology
University of Calcutta
Kolkata
West Bengal
India

ISSN 2191-5407   ISSN 2191-5415 (electronic)
ISBN 978-81-322-2174-6 ISBN 978-81-322-2175-3 (eBook)
DOI 10.1007/978-81-322-2175-3

Library of Congress Control Number: 2014953933

Springer New Delhi Heidelberg New York Dordrecht London

Printed on acid-free paper

Springer is part of Springer Science+Business Media (www.springer.com)

For our dear Ahona

Foreword

Over the past decade, researchers have extensively used various types of nanoparticles for the development of advanced therapeutic tools by exploiting the advantages that nanometer-size particles offer. This book covers recent developments on use of various types of nanoparticles in lung cancer therapy. The authors concentrated on the interaction mechanism of nanoparticles with the human body and how we can avoid the toxic effect of nanoparticles by using the design rules found in nature. This book will be ideal for undergraduate and postgraduate students who are interested in this exciting field of research, because this book will guide them to develop functional nanoparticles to enhance therapeutic index of nanoparticles-based drug/gene/imaging probe delivery vehicles without causing damage to the human internal system.

Prof. Suprakas Sinha Ray
Chief Research Scientist and Director
DST/CSIR Nanotechnology Innovation Centre
National Centre for Nano-structured Materials
Council for Scientific and Industrial Research, Pretoria
South Africa

Preface

This book presents a detailed study on efficacious therapeutic values of various organic and inorganic nanoparticles on lung cancer diseases. In clinics, different surgical and non-surgical therapeutic approaches including chemotherapy, photo-dynamic therapy, and radiotherapy are employed in the treatment of lung cancer. However, most of the techniques lack the knowledge of differentiation between healthy and abnormal cells with concomitant loss of life. Even chemotherapeutic drug solubility is limited in the human bloodstream and hardly reaches the rigid and viscous zone of solid tumors. To overcome this problem after a decade of multidisciplinary research, scientists have been able to synthesize target-specific natural or synthetic nanoparticles to diagnose lung cancer. Due to its very small size and high invasive property they can invade tumor cells and perform multidisciplinary tasks ranging from prognosis of cancer, early detection of cancer cells, cellular imaging, and localized drug delivery to treatment of cancer. So far, very limited research has been done in the development of nanoparticles, concentrated only on lung cancer diagnosis. One primary reason may be due to non-availability of binding ligands on nanoparticles to target specifically to lung cancer cells. However, in the present decade with the development of ligands like aptamers, anisamide, anti EGF, and siRNA the problem of specific targeting has been circumvented and various nanoparticles have been tuned to enable lung cancer diagnosis at an early stage. Here we present how different lung cancer cells targeting antibody functionalized nanoparticles can cure the so difficult to treat lung cancer diseases easily at an early stage and offer a healthy prolonged life to affected people. This book will be a hope to young researchers developing modified nanoparticles in an attempt to open up newer and cheaper diagnosis tools for lung cancer patients; here, we have specifically elaborated suitable characteristic properties and mechanistic details of each of the diagnostic nanoparticles in the treatment process; a guidance to doctors who clinically try these newly developed techniques for faster and better treatment; and an information kit to patients and their families for better understanding of the nanotechnology-based lung cancer treatment processes so that more and more of them could welcome these new medical field without much fear.

Acknowledgments

Our deepest thanks:

To all the scientists working day and night to develop novel techniques by clubbing already available treatments and nanotechnology in order to treat lung cancer at an early stage. Their extensive research and effort to commercialize a few of the formulations has motivated us to create awareness among the general public and among doctors to welcome new medical facilities based on nano-biotechnology.

To the publisher, Springer (India) Pvt. Ltd. for giving us this wonderful opportunity to share our knowledge with the entire world.

To our lab members Dr. Tridib Bhunia, Mr. Arindam Giri, Mr. Abhijit Pal, Ms. Srijani Sengupta and Mr. Soumen Sardar for their constant inspiration in writing this book.

To our parents and beloved ones.

Contents

About the Authors

Dr. Abhijit Bandyopadhyay Assistant Professor, Department of Polymer Science and Technology, University of Calcutta, Kolkata, India.

Dr. Bandyopadhyay is M.Tech. and Ph.D. in Polymer Science and Technology and is presently working as Assistant Professor in the Department of Polymer Science and Technology, University of Calcutta and Technical Director in the Board of Directors of the upcoming South Asian Rubber & Polymers Park in West Bengal, India. He is former Assistant Professor of Rubber Technology Centre at Indian Institute of Technology, Kharagpur. He has more than 8 years of Teaching and Research Experience and has published more than 60 research papers in high impact international journals, four book chapters and received one Indian Patent. He has guided two Ph.Ds and presently eight students are working under his supervision. He has received awards like Young Scientist Award from Materials Research Society of India, Calcutta Chapter, Young Scientist Award from Department of Science and Technology, Government of India and Career Award for Young Teachers from All India Council for Technical Education, Government of India. He has delivered many Invited Lectures in international conferences both in India and abroad. He is a life member of Society for Polymer Science, India and the Fellow member of International Congress for Environmental Research.

Tamalika Das Department of Polymer Science and Technology, University of Calcutta, Kolkata, India.

Ms. Tamalika Das was born in Kolkata in the year 1986. She was educated at St. Xavier's Institution (Kolkata) and graduated (2008) in Chemistry (Hons.) from Scottish Church College (Kolkata). She completed post B.Sc., B.Tech. (2011) and M.Tech. (2013) from University of Calcutta in the Department of Polymer Science and Technology. She earned gold medal twice from University of Calcutta (both in B.Tech. and

M.Tech.). She did her M.Tech. project from the Indian Association for the Cultivation of Science (Kolkata). Currently, she is pursuing doctoral research in the Department of Polymer Science and Technology at University of Calcutta. Her area of interest is hyperbranched polymers. She has published one paper in high impact international journal so far. She enjoys INSPIRE Fellowship from Department of Science and Technology, Government of India.

Sabina Yeasmin Department of Polymer Science and Technology, University of Calcutta, Kolkata, India.

Ms. Sabina Yeasmin was born in 1986 in a small village in Burdwan, West Bengal, India. She got her education from Kalna Hindu Girls High School. She received Bachelor of Science degree in Chemistry from Maulana Azad College (Kolkata) and Master's of Science in Biochemistry from West Bengal State University (Kolkata). She stood first in the Post Graduate Examination and received INSPIRE fellowship from Department of Science and Technology, Government of India. Presently, Ms. Yeasmin is pursuing her doctoral degree in the Department of Polymer Science and Technology of University of Calcutta in the area of synthesis, characterization and biomedical applications of silver nanoparticles.

About the Book

This brief provides an insight into the present scenario of the role of nanotechnology in the diagnosis and treatment of lung cancer at an early stage. Currently, lung cancer is a subject of major concern owing to the very high mortality rate throughout the world. Most of the conventional treatment methods such as surgery, chemotherapy, radiotherapy, etc., fail to prolong the life of the patients. Incidents of recurrence are also very common in case of lung cancer. Researchers have shown that nanoparticles may act as a powerful anticancer tool, especially for lung cancer. Unique surface properties and easy surface functionalization of nanoparticles enable early detection, diagnosis, imaging, and treatment of lung cancer. The authors have elaborately presented how various nanoparticles (natural, semi-synthetic, and synthetic) may help in the treatment of lung cancer. They have also detailed works of various scientists who succeeded in developing effective nanoparticles and enabled very specific lung cancer therapy without any undesirable side effects and minimized death.

Chapter 1
Introduction

Abstract Integration of nanotechnology into the field of medical science demands detailed understanding of the physicochemical properties of various nanoparticles and their interactions in the biological systems. Lung cancer being one of the deadliest diseases in the world with very high mortality rate may be cured with combination pack of molecular therapy, conventional therapy and nanotechnology. However, any wrong treatment may lead to death of the patients and may obstruct the emergence of nanotechnology in medical field. Hence, we need to understand the anatomy of human lung, reasons for lung cancer, highly affected parts in lung cancer, stages of lung cancer, rate of metastasis to other parts, available treatments and drawbacks in the conventional treatment methods, which drive the necessity for including nanotechnology in the diagnosis and treatment for lung cancer. Now that introduction of nanoparticles into human body may be an ethical issue. Doctors and patients may not gladly accept the rising technology. Hence, here we tried to brief how nanoparticles work in human system and few possible threats associated with the use of nanoparticles which may be wisely overcome by precise design of nanoparticles intended for the treatment of specific biological disease via a specific way.

Keywords Lung cancer · Metastasis · Therapy · Nanoparticle (NP) · Specific biodistribution

1.1 Cancer

Cancer is one of the scariest diseases. It is often thought that cancer is untreatable and patients die a very painful death. Nevertheless, this deadliest story of cancer is well exaggerated due to lack of proper knowledge on the disease and its possible treatment. Undoubtedly, cancer is a very serious disease which takes away millions of life every year. According to a report by the Cancer Research Organization (UK), in the year 2012, nearly 14.1 million people around the world were diagnosed with cancer and 8.9 million adults were dead because of cancer as represented in Fig. 1.1 [1].

© The Author(s) 2015
A. Bandyopadhyay et al., *Nanoparticles in Lung Cancer Therapy - Recent Trends*,
SpringerBriefs in Molecular Science, DOI 10.1007/978-81-322-2175-3_1

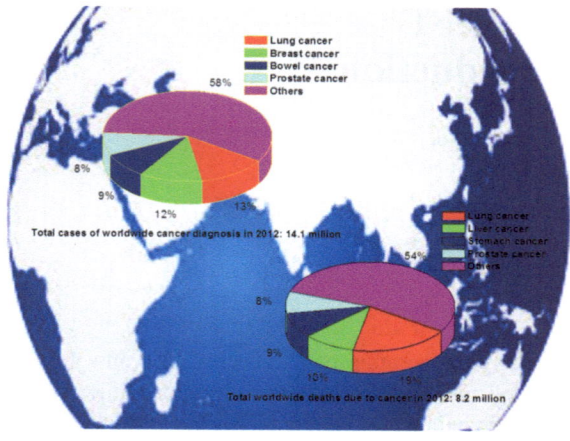

Fig. 1.1 Statistics of the most common cancers resulting in death around the world in the year 2012

With the advancement in medical technology, most form of cancer is curable, seemed to extend patients' life by many years and even alleviating pain and sufferings of their last days. However, patients' lack of interest in accepting new cancer treatment processes has been the primary reason for increment in cancer mortality rate throughout the world. Hence, it is necessary to educate cancer patients and their families on the basis of cancer causes, growth, spreading and available modern cancer cure techniques.

Any form of cancer is defined as the growth of cancerous cells that divide without any control and have the ability to infiltrate and to destroy normal tissues. Generally, human body maintains a well-controlled system that puts a check on the cell growth after a certain age (depending upon body structure and requirement) so that new cells are produced only when needed and subsequently replace defective or dead cells. Imbalance in the cell cycle owing to mutation in certain genes of cells (called oncogenes, for example, growth-promoting genes for the signaling protein Ras become highly active) results in uncontrolled division and proliferation of harmful cells that eventually form a mass known as neoplasms or tumor (Fig. 1.2) [2]. Mutation also inactivates the tumor suppressor gene and the DNA repair gene (p53, a multifunctional protein that normally senses DNA damage and induces apoptosis or cell death). Thus, any growing tumor may be fatal as they take up the oxygen and other vital nutrients from the neighboring cells and force them to behave abnormally, i.e., generates other sites of emerging tumors. Invasive cancer cells also produce enzyme proteases, which subsequently degrade the extracellular matrix of the adjacent tissues and hence gain access to new territories. Moreover, due to loss in cell adhesion of cancerous cells with other cells, they freely traverse to other parts of the body through the blood stream and the lymphatic systems and develop secondary tumors.

Tumors may be benign or malignant. The term "cancer" refers to those tumors that are malignant. Benign tumors can be eradicated from the body without affecting the normal tissues. They have slow rate of proliferation, are encapsulated, and

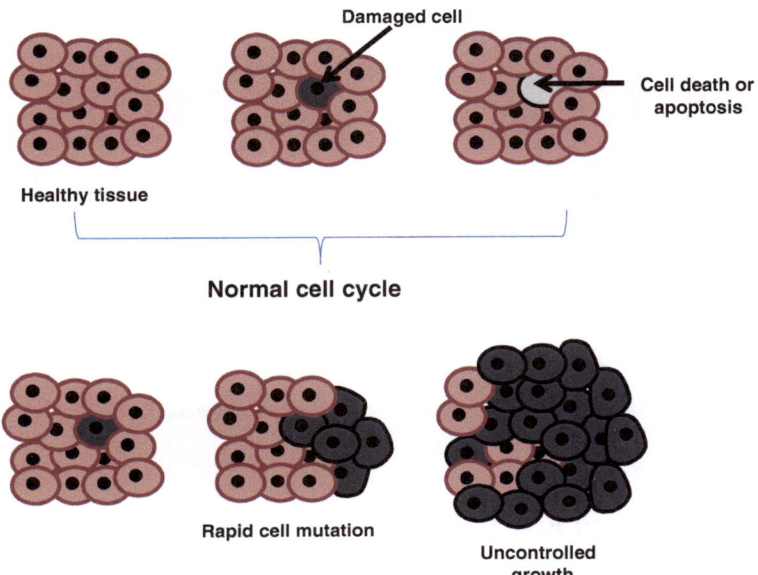

Fig. 1.2 Normal cell cycle and uncontrolled division in cells leading to formation of tumors

do not infiltrate surrounding tissues. Death does not occur due to such benign tumors. On the contrary, malignant tissues grow irregularly, have a high rate of cell proliferation, invade healthy tissues (lack differentiation), are resistant to apoptosis, and then migrate to the bloodstream, lymphatic system, or other parts of the body. This process of spread is known as "metastasis" (literally meaning "new places") [2]. It is reported that some type of metastasis cancer may be cured with the current treatments available, but most cannot be. In general, the primary goal of this treatment is to control the growth of these cancer cells and to remove the symptoms caused by it. This helps to prolong the patient's life. In fact, most patients who die of cancer actually die of metastatic cancer. Virtually all cancers, including cancers of blood and lymphatic system, are metastatic in nature. The most common sites of metastasis are lungs, bones, and liver.

There are approximately 200 types of cancer, each with different causes and symptoms. An individual's risk of developing cancer depends on many factors, including age, lifestyle, and genetic makeup. The most common causes of death due to cancer is lung cancer as reported from a statistics study report by UK cancer research group in the year 2011 (Fig. 1.3). The main reason behind such a high lung cancer mortality rate is obviously lack of proper knowledge on estimating the spread of cancer to different vital organs at an early stage and inappropriate delivery of exact dosage of anticancer drugs only to specific sites. Hence, here we tried to give a detailed account of the disease of lung cancer and recent ongoing research/clinically tried processes for its alleviation, focusing mainly on the use of nanoparticles in therapy.

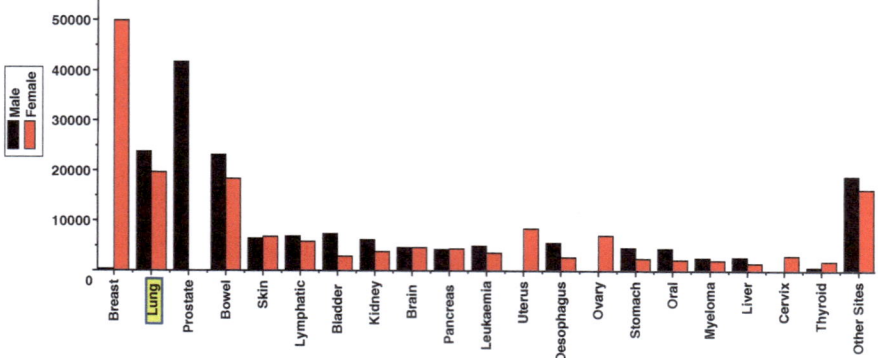

Fig. 1.3 Statistics of the 20 most common causes of death from cancer in UK in the year 2011 (reproduced) [3]

1.2 Lung Cancer: Few Facts and Measures

Lung cancer is the deadliest type of cancer among both men and women. However, it is more common in people above 45 years. Human lung is a very common site for metastasis from tumors growing in other parts of the body [2]. We know the primary function of the lungs is to exchange gases between the air we breathe in and the blood. Through the lungs, carbon dioxide is expelled from the bloodstream and oxygen is inspired into the bloodstream. 90–95 % lung cancer is thought to arise from the epithelial cells, lining the bronchi and bronchioles (major airways entering the lungs). Lung cancer, also known as "bronchogenic carcinomas," is widely classified into four major classes by international standards (WHO 1982), based on the histological appearance: adenocarcinoma (AD), squamous cell carcinoma (SCC), large cell lung carcinoma (LC), and small cell lung cancer (SCLC). Since the origin, metastatic behavior, prognosis, and treatment for SCLCs are markedly different from NSCLCs, the four histological subtypes are clinically divided into SCLC and non-small cell lung cancer (NSCLC) comprising of AD, SCC, LC, and other minor forms [4]. SCLCs are very aggressive in nature and occur mainly due to extensive tobacco smoking, persistent inhalation of asbestos, radon gas, gasoline or diesel exhaust, or consumption of high levels of arsenic in drinking water. SCLCs originate from an inner layer of the walls of bronchi (bronchial submucosa) and metastasize very rapidly to many sites within the body (Fig. 1.4) but is generally detected at the later stage when it has spread extensively which is only a few months if left untreated. While talking about lung cancer, it is essential to focus on "pancoast tumors". These are bronchogenic tumors occurring at the lungs apex (also known as superior pulmonary sulcus). The name derives from its location in the lungs. It rapidly spreads to nearby tissues of ribs and vertebrae, i.e., extensively metastatic in nature. In fact, they are categorized

Fig. 1.4 Spread of malignant tumors (SCLCs) throughout the lung (*Credit* image courtesy, http://yes-give-up.blogspot.in/p/give-up-smoking.html. Access date 20 Nov 2013)

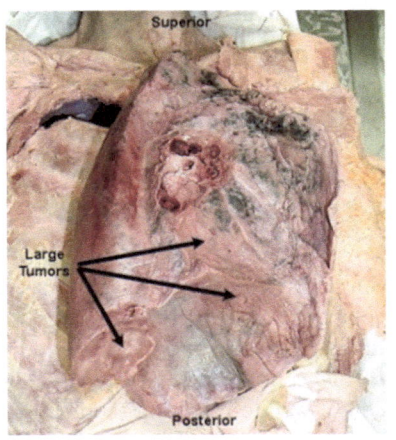

as SCLCs. These are very difficult to cure as they are reluctant to respond to the conventional therapies. However, they account for only 20 % (ACS 2010). NSCLCs account for about 80 % of all lung cancer and spread slowly to other body parts. Hence, NSCLCs are curable.

It is very difficult, in fact practically impossible to detect lung cancer at an early stage unlike other forms of cancer as there are few symptoms for which the patients would worry about and consult doctors. However, some of the early apparent signs of lung cancer include persistent coughing accompanied by hoarse throat, chest pain, reduced appetite, unintentional weight loss, respiratory infections, and difficulty in breathing. Undoubtedly, patients must be aware of the symptoms at the preliminary stage and immediately go through a thorough checkup. As the cancer spreads, patients may experience severe symptoms such as lack of concentration, persistent coughing with blood laden sputum, pains in various parts of the body, feeling of fatigue/dizziness all the time, difficulty in swallowing, and jaundice. Well in both the early and late stages of lung cancer (if suspected from the symptoms), patients must try to be safe than to be sorry.

1.3   Limitations of Conventional Treatments for Lung Cancer

Metastatic cancer including those of lung cancer may be treated with systematic therapy (chemotherapy, biological therapy, targeted therapy, and hormonal therapy), local therapy (surgery, radiation therapy), or a combination of these treatments. The choice of treatments depends upon the type of primary cancer (size, location, and the no. of metastatic tumors), patient's health and age, and obviously the stage of cancer. Therapy may be curative (eradication possible) or palliative (measures that are unable to cure cancer but can reduce pain and suffering) [2]. It is reported that more than one type of therapy is prescribed to enhance the effects of the primary therapy and is referred to as adjuvant therapy. Examples of such therapy include chemotherapy or radiotherapy, which are administered after surgical removal of tumors to destroy any remaining unwanted cells.

1.3.1 Surgery

Surgical removal of the tumors (Fig. 1.5) is generally performed for stage I or sometimes stage II of NSCLCs, provided the cancer has not spread beyond the lungs. Some commonly employed surgical techniques for the removal of lung cancerous cells include lobectomy (surgical excision of lung lobes containing cancer cells), wedge resection (surgical removal of a section/mass of harmful tissues of lung lobe and not the entire lung lobe), and pneumonectomy (removal of the entire lung). It is not recommended in cases of large tumor size, locations close to vital organs, and the presence of distant metastases [5]. About 10–35 % cancer may be removed, but it is never possible to eradicate the damaged cells. A certain percentage persists or recurs even after surgery. However, surgery is restricted to patients having serious heart or lung diseases as they may not survive the operation. Surgery is less common with SCLCs as these tumors are less likely to be localized to one area to be operated. There are other shortcomings of surgery. Following the surgery, patients may suffer from difficulty in breathing, pain, and weakness. There may be risks of bleeding, infection, and other complications.

1.3.2 Chemotherapy

As already mentioned that surgical approaches are not useful for patients suffering from lung cancer at a late stage, hence more radical treatment methods (chemotherapy or radiation therapy) are practiced for severe cases. Chemotherapy is basically the administration of drugs into the patients' bodies, either after surgery or just solely to prevent the growth of cancerous cells either by killing them or by obstructing the

Fig. 1.5 Surgical removal of tumors from lung (*Credit* image courtesy by Kathy Jones, http://www.medindia. net/news/Adverse-Quality-of-Life-After-Esophageal-Cancer-Surgery-may-be-a-Worrying-Sign-95836-1.htm. Access date 20 Nov 2013)

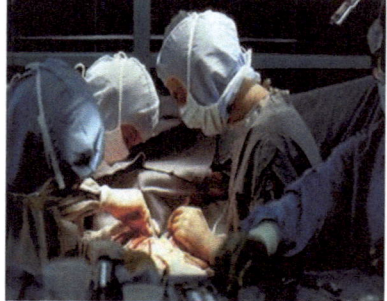

cell division processes. It may reduce the symptoms and sufferings associated with cancer. Both SCLCs and NSCLCs may be treated with chemotherapy. Interestingly, chemotherapy prolongs life of the patients suffering from SCLCs by four- to five-folds. Chemotherapy may be given as pills or intravenous infusion or a combination of these two (Fig. 1.6a). The basic process of chemotherapy relies on enhanced permeation and retention effect (EPR effect; passive targeting) of tumors (Fig. 1.6b). Tumors exhibit hyperpermeability owing to upregulation in vascular endothelial growth factors (VEGFs), nitric oxide (NO), bradykinin peptides, prostaglandins, collagenases, etc. Thus, tumor-targeting macromolecules (proteins, RNA, DNA, etc.) may easily target and diffuse through the leaky tumor vasculatures. On the contrary, tumors also exhibit high retention ability of any foreign molecules. In the tumors' interstitial places, high interstitial pressure (due to emergence of additional blood vessels; "angiogenesis" as discussed broadly in Sect. 4.1) and lack of proper lymphatic network for drainage of foreign materials from cellular components cause accumulation of the invaded macromolecules within the tumors as compared to other healthy tissues. Thus, by the EPR effect of tumors, most of the chemotherapeutic drugs stay for a longer time to effectively kill the host cancerous cells. Mostly, the platinum-based drugs are effective in the treatment for lung cancers by chemotherapy. Commonly employed platinum based anti cancer drugs includes cisplatin (cis-diamminedichloroplatinum (II), CDDP) [77] and carboplatin [cis-diammine (1,1-cyclobutanedicarboxylato) platinum (II)]. Also, paclitaxel (taxol; PTX) is another potential chemotherapeutic drug worthy to mention which is used to treat patients who cannot tolerate platinum compounds. However, taxols are often used in conjugation with cisplatins and carboplatins in the treatment for lung cancer for synergistic effects. Chemotherapy may be unsuccessful in the treatment for tumors with poor vasculatures and low surrounding interstitial pressures. Chemotherapeutic drugs work more effectively at the peripheral region of tumors

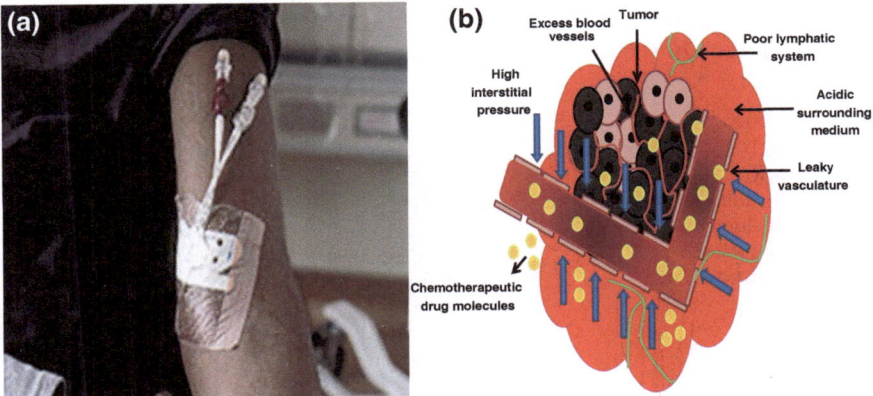

Fig. 1.6 a Administration of drugs into patient's body during chemotherapy. **b** Accumulation of chemotherapeutic drug molecules in tumor by EPR effect caused by tumor microenvironment (*Credit* for Fig. 1.6a: Image courtesy by National Cancer Institute, https://visualsonline.cancer.gov/details.cfm?imageid=4489. Access date 21 Nov 2013)

where EPR effect is high as compared to the central part of the tumors where drug extravasations are reduced. Hence, tumors are not entirely treated and cases of recurrence are common phenomenon. The major problems of chemotherapy are that the drugs used may kill the normal rapidly dividing cells (similar to tumors) such as those of gastrointestinal tract, bone marrow cells, and hair follicles, resulting in unpleasant side effects and even death [5]. Often due to poor vasculatures in tumor microenvironments, oxygen is not readily supplied and carbon dioxide gets accumulated in tumors which generate lactic acid by anaerobic glycolysis. Hence, chemotherapeutic drugs get ionized and are prevented from extravasations into tumors. Even due to biochemical and metabolic changes in cancerous cells (which results in upregulations in multi drug resistant proteins called P-glycoproteins, alterations in enzyme activities, intra- and extracellular transport mechanisms, cell cycle, etc.), tumors develop strong resistance against chemotherapeutic drugs. Drug resistance by certain tumors is thus of another potential obstruction in the working of conventional chemotherapeutic drugs. Often platinum-based chemotherapeutic drugs cause nephro- and cardiotoxicity. Other side effects include fatigue, gastrointestinal distress, low white blood cell count, anemia, weight loss, hair loss, nausea, vomiting, diarrhea, and mouth sores.. However, the side effects disappear during recovery or after the completion of the treatment.

1.3.3 Radiation Therapy

Radiation therapy or radiotherapy may be employed as a treatment basis for both SCLCs and NSCLCs (Fig. 1.7). It uses high-energy X-rays or ionizing radiation {cobalt (^{60}Co), radium (^{228}Ra), iodine (^{131}I), radon (^{221}Rn), cesium (^{137}Cs), phosphorus (^{32}P), gold (^{198}Au), iridium (^{192}Ir), and yttrium (^{90}Y)} to kill the cancer cells [6]. This type of therapy only shrinks a tumor or limits its growth by directly damaging DNA or by generating reactive oxygen species (ROS), even when rendered as a sole therapy [5]. Yet it is a common method of treatment, if the patient refuses for surgery,

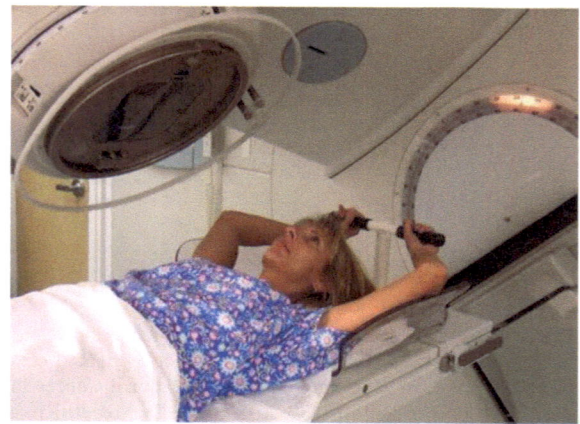

Fig. 1.7 Image of radiotherapy (*Credit* image courtesy by Radiation Oncology Health Care, http ://www.rohbaltimore.com/ Diease-Information/Breast-Cancer.php. Access date 21 Nov 2013)

or if the tumors have spread irregularly to various sites or if there is threat to life on operation. Generally, the radiation therapy is based on focusing a tumor by radiation from all directions. It is of two forms: (a) brachytherapy—where the radioactive source in pellets is placed close to the tumor and (b) teletherapy—where the patient is irradiated from a source placed outside the body. It lacks the risks of major surgery [5]. However, it may be combined with surgery to bring out better results. This is generally used before surgery to shrink tumors or after surgery to kill the remaining harmful cells. Radiation therapy may be applied as a curative or a palliative or as an adjuvant therapy in combination with chemotherapy, depending upon seriousness of the diseases as judged by the doctors. Yet it has certain side effects such as fatigue, reduced white blood cell count, and low blood platelet levels. If the digestive organs (radio sensitive) are on the way to radiation, patients may suffer from nausea, vomiting, diarrhea, and skin irritation. Another drawback in radiation therapy is the development of radiation resistant tumors due to the expression of ROS scavengers.

1.3.4 Photodynamic Therapy

In photodynamic therapy (PDT), an organic photosensitive agent (e.g., porphyrin, a naturally occurring substance, texaphyrin, chlorine, etc.) is injected into the patient's body, which is rapidly picked up by the cancer cells, even at various stages of cancer. Porfimer sodium (a first generation photosensitizer) is often used in the treatment for early and later stage of lung cancer. When sufficient amount of photosensitizers accumulate in the affected sites, then in the presence of external light source of certain wavelength, they get excited from the ground state to excited singlet and later to triplet states. Again, molecular oxygen species that are abundantly present in tissues stay at ground triplet states. Hence, an energy transfer occurs between the excited photosensitizers and molecular oxygen species whereby excited singlet states of oxygen molecules are generated. These singlet oxygen species are highly reactive and thus selectively destroy the cancer cells. This treatment is useful in case of inoperable lung tumors (say when tumors are hidden or undetectable in chest X-rays) and is very effective as it lacks non-specificity [2]. Yet in this type of therapy, patients may suffer from inflammation and difficulty in breathing or swallowing. PDT being a very effective anticancer treatment tool, research has produced many PDT agents. However, most of the PDT agents are scarcely soluble in aqueous medium. Hence, delivery of PDT agents to tumors is a challenge and requires effective carrier medium for their successful accumulation in specific locations.

1.3.5 Treatment of Recurrence

Cancer may elapse after surgery, chemotherapy, and/or radiation therapy. If the recurrent cancer is confined to a single site, then it can be treated with surgery or

PDT. However, in most of the cases of recurrence, patients do not survive when treated with conventional therapeutic methods.

1.3.6 Experimental Therapies

Worldwide different cancer therapies fail most of the time. So, numerous experimental therapies are tried by the doctors in clinics on the patients. However, due to the lack of information on the outcome of this type of treatment, patients and their families have to be strong enough mentally for any extremes. Nowadays targeted therapies have emerged in alternative to the critical treatments discussed so far. The drugs erlotinib (Tarceva), gefitinib (Iressa), and cetuximab are mainly used as targeted drugs in curing NSCLCs, especially when patients do not respond to chemotherapy or radiation therapy or other adjuvants [2]. Targeting is done to proteins (epidermal growth factor receptors) growing at abnormally high levels on the surface of the damaged cells simply by conjugating drug molecules or imaging probes with monoclonal antibodies. However, most of the targeted macromolecules fail to reach the destination and get cleared through the renal system. Other targeting therapies include blocking the newly developed blood vessels within a cancer with a targeted drug called antiangiogenesis. Hereby, essential nutrients and oxygen are blocked to cancer cells and they die. No doubt that these targeted therapies are of potential interest in the treatment for any form of cancer. So, manipulation of targeting is essential. One of the most welcomed technologies in this field is the "nanobiotechnology", more specifically narrowing to "nanomedicine". However, most of the cancer therapeutic strategies as mentioned above are effective only at the initial stages of NSCLCs and hardly in case of SCLCs, which mostly rely on surgery or gene therapy. Owing to the potential threat to human life from lung cancer, here we made a detailed study on lung cancer diagnosis using variety of natural/synthetic, metallic/non-metallic, and polymeric nanoparticles.

1.4 Nanotechnology—A New Promising Therapeutic Tool

Nanotechnology covers a huge area of applications starting from cosmetics, packaging, and textiles to electronics, optical devices, solar cells, etc. Today nanotechnology plays a pivotal role in the detection, diagnosis, imaging, and targeting of lung cancer cells. Issac Asimov in his book "Fantastic Voyage" told a story of five persons who were reduced to microscopic size and then sent in a subatomic submarine through a dying man's carotid artery for the treatment of a blood clot in his brain. Interestingly, such concept of miniature particles in a science fiction led to the success story of nanotechnology in the medical field through extensive research. The National Institute of Health (NIH, USA) defined nanomedicine as "an offshoot of nanotechnology, which refers to highly specific medical intervention at the molecular scale for curing disease or repairing damaged tissues, such as bone, muscle or nerve." Nanotechnology put a marked effect in cellular imaging, diagnosis, experimental therapies, etc. Undoubtedly nanotechnology has also intervened into the realm of diagnosis and treatment for lung cancer in an attempt to reduce mortality rate. Firstly, it is essential to explain what difference a nanoparticle (NP) makes in contrast to a conventional macroparticle in any cancer diagnosis. By definition, particles having at least one dimension in the nanometer scale (especially less than 100 nm) are termed as nanoparticles (NPs) (Fig. 1.8) [7]. NPs have attracted significant attraction in nanobiomedical technology due to the presence of large proportions of surface particles with high energy with respect to the bulk of the material and the quantum confinement effect (the electrons are squeezed into a small area), which enable them to adsorb and carry both hydrophilic and hydrophobic macromolecules (DNA, RNA, drugs, probes, etc.) even through the complicated barriers of human circulatory system, under different body environments. Being small in size with high surface area, NPs may easily enter the cancerous cells if properly targeted and improve cancer detection and treatment processes. In fact, nanoscale devices and components are of the same size as biological entities, however, smaller than the human cells [8]. It appears that nature, in making the biological systems, has extensively used nanometer scale. If one has to go hand in hand with nature in treating any cellular diseases such as cancer, one needs to use the same scale [9]. NPs are smaller in size than human cells (10,000–20,000 nm in diameter) and organelles. However, they are similar in size to that of the biological macromolecules such as enzymes and receptors (e.g., hemoglobin and lipid bilayer surrounding the cells are typically 5 and 6 nm, respectively). The extremely small size of the NPs (carrier of drugs) opens the pathway through the various biological barriers or tight junctions within the body, into the cell and various cellular compartments including the nucleus [10]. Nanoscale devices smaller than 50 nm may easily pass through most of the cells. However, those below 20 nm may transit out of the blood vessels and exit the body, thus of no use. Hence, NPs having particle size ranging from 50 to 100 nm are considered as efficient anticancer tools. NPs get uniformly distributed throughout a human body. In fact, the biodistribution of NPs is rapid (within half an hour to 3 h) [10]. However, there are numerous locations where NPs may not reach easily. By understanding the size and surface property requirements for reaching specified sites within the body, localization of NPs to these sites may be accomplished. Basically, localization and targeting of NPs is done by decorating their surfaces with special targeting ligands that provide NPs–cell interactions and drastically influence the final location of the NPs in the body. For example, an addition of a moiety such as a small molecule (peptide, protein, or antibody) to the NPs may target the latter to the damaged cells [11]. Using NPs, it may also be possible to achieve improved delivery of poorly water-soluble drugs or imaging probes. Attachment of drug or probe molecules on the surface of small-sized particles increases the total surface area of the drugs allowing faster dissolution in the blood stream, which contains 70 % water. Such drug or probe molecules with improved aqueous solubility are absorbed faster by human body and thus delivered at specific cells or tissue locations [12]. Again, many of the orally or intravenously administered anticancer drugs or probes suffer from major shortcomings, which include modification or degradation of the drug/probe molecules in the acidic environment of stomach, alteration in structure of drug/probe

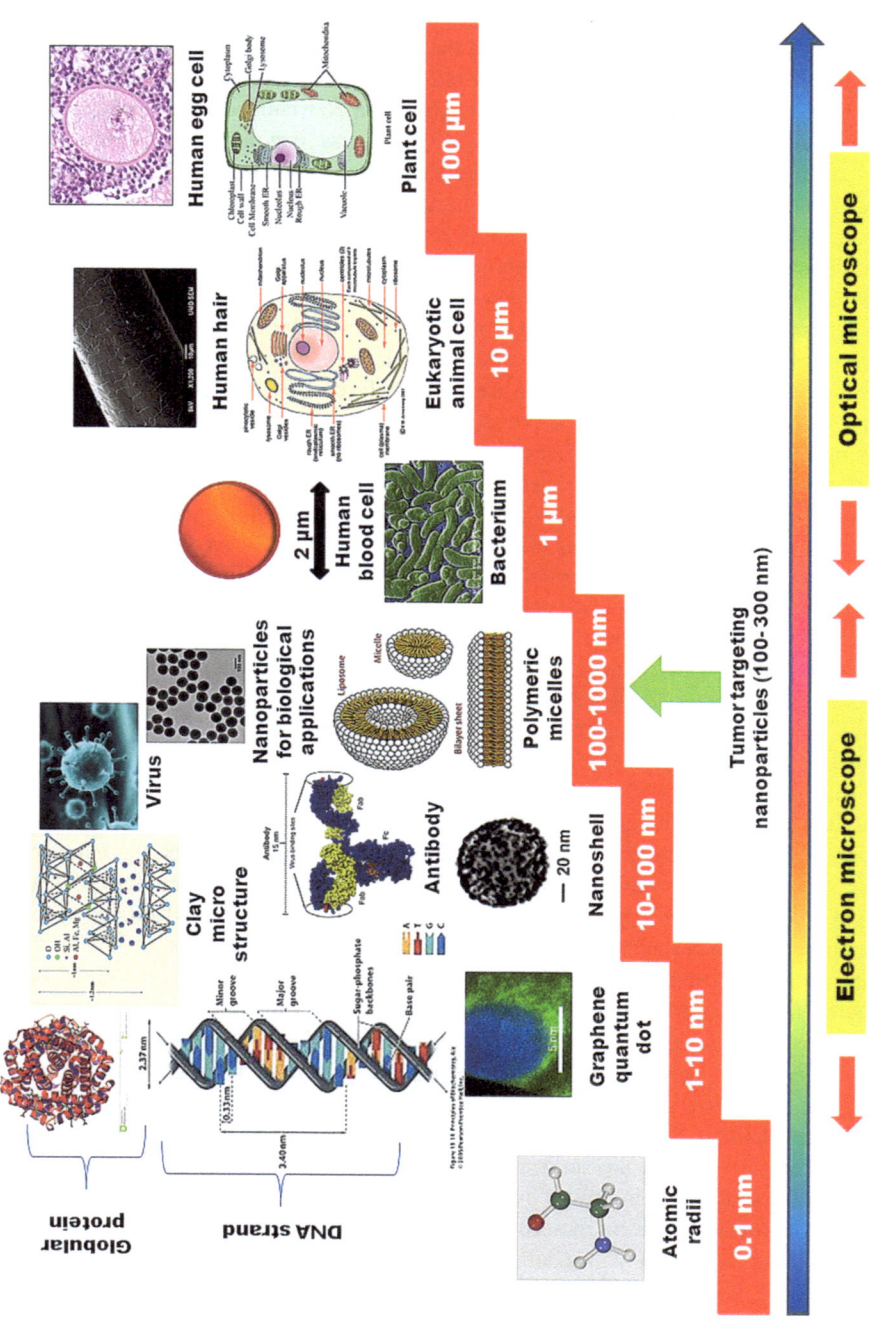

Fig. 1.8   Pictorial representation of various natural and synthetic components in different size ranges

molecules due to metabolism by the liver while circulating through the hepatic portal system (first pass effect), resistance to drug by the intestine resulting in low absorption of drugs, lack of targeting to damaged cells, drug overdose, frequent administration, and thus premature loss of life. NPs may be useful in overcoming such shortcomings of conventional therapeutic agent delivery methods. Tumor microenvironment is highly acidic than those of the normal cells. pH-sensitive NPs loaded with active drug molecules or imaging probes, if designed to be stable at physiological pH (6.0–7.0) and unstable at low pH, may successfully release active macromolecules in the target cells. However, in general, NPs are stable over a wide range of temperature and pH. Hence, therapeutic properties of NPs are tuned by specific functionalization on their surfaces. Even NPs being biocompatible (for short-term presence in the circulatory systems) may be easily administered through various routes. Hence, NPs find potential applications in the realm of drug and probe delivery. It is obvious that if the drugs and imaging probes are loaded with nanocarriers, then there would be significant enhancement in diagnosis and therapy, respectively. Even some drugs that have failed clinical tests earlier could be reapplied using nanotechnological approaches. Interestingly, pharmaceutical sciences are using NPs to reduce toxicity and side effects of various drugs: i.e., enhance specific biodistribution and targeting in sufficient concentrations and efficiently to heterogeneous population of damaged cells and tissues to exert the pharmacological effect without causing irreversible unwanted injury to healthy tissues and cells, over a certain period of time. Actually, conventional approaches do not effectively differentiate between damaged and normal cells, i.e., lack specificity. This normally causes system toxicity with severe and adverse side effects and concomitant loss of quality of life. NPs improve bioavailability which refers to the presence of drug molecules where they are needed in the body (i.e., in the area with chronic imbalances and deficiencies) and where they will do most good (Fig. 1.9) [9]. This is referred to as "targeted drug delivery." Basically,

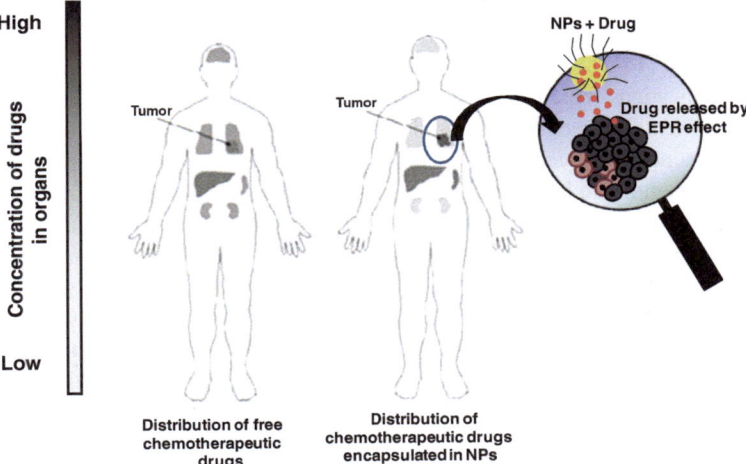

Fig. 1.9 Concentrated accumulation of NPs-based drug delivery carriers in lung tumors as compared to free drug molecules

targeted drug delivery, sometimes called "smart drug delivery," is a method of delivering medication to a patient in a manner that increases the concentration of the medication in some parts of the body relative to others, in a "sustained" or "zero-order (drug release occurs at a constant rate and is independent of the concentration of drug)" dosage form for a prolonged time. Targeted release systems reduce frequency of the dosages taken by the patient, having a more uniform effect of the drug, reduction of drug side effects, and reduced fluctuation in circulating drug levels. Also, they extend the drug systemic circulation lifetime. In traditional treatment processes, chemotherapeutic drugs are administered to cancer patients in maximum tolerable dosages. This obviously causes high system toxicity. Hence, patients are given rest periods between cycles of therapies for the dilution of the chemotherapeutic drug molecules in unwanted parts of the body. Rest periods disturb the continuity of therapy as tumors do not receive drug all the time and regrowth of tumors happen. Even sometimes newly grown tumors are more malignant in nature and are highly resistant to traditional chemotherapeutic drugs which were already used in the system. In targeted therapy using NPs, limited dosage of drug molecules is administered to patients' bodies as NPs deliver the macromolecules in a sustained fashion and the rest get cleared from the body. However, although a sustained release of strong chemotherapeutic drugs reduces system toxicity from overdose, yet a potential threat to toxicity develops due to extremely small size of the NPs. Nothing is perfect on this earth. NPs too suffer from various shortcomings in the biological systems. NPs may easily roam throughout the body and may get accumulated in biological cells or organelles. The problem is more pronounced with inorganic NPs which are bioincompatible and thus get accumulated in the body, exerting undesirable side effects such as poison. Sometimes NPs on entering the cells may induce oxidative stress with the generation of ROS, which causes inflammation and even DNA damage [8]. However, "stealth coating" of the NPs with biocompatible surface modifiers such as polyethylene (PEG) may transform them to a biocompatible entity, minimize degradation [13], impart non-immunogenicity, and also prolong their life span in the circulatory system by reducing renal ultrafiltration owing to an increase in effective molecular size of the particles [14–16]. Hence, any NPs if designed for in vivo (Latin for "within the living") applications must clearly be verified that they do not possess nanotoxicology and environmental side effects. Often the efficiency of biological labels on the surface of NPs is potentially reduced in the harsh body environment. The problem is more pronounced in asymmetric biomolecules such as antibodies [17]. However, the problem of losing antigen binding capacity could be overcome by cite specific binding of antibodies onto the surface of biosensors via physicochemical adsorption and covalent attachment [18]. Again, biodistribution of NPs in human body is sometimes affected by undesirable interactions with biological molecules and processes such as opsonization (deposition of blood opsonic factors such as immunoglobulins on the surface of NPs which facilitates their binding with phagocytic cells) [8]. Hence, most of the therapeutic NPs are cleared by mononuclear phagocytic systems (which consists of monocytes or leukocytes in the bloodstream and macrophages in the liver or spleen that take up and metabolize foreign molecules and particulates). In certain circumstances, another potential trouble caused by circulation of NPs in human body is hemolysis. NPs may induce hemolysis owing to adsorption of hemoglobin on NPs. This is similar to opsonization.

Hence, opsonized NPs are easily recognized by phagocytic cells and rapidly cleared by the reticular endothelial system (RES). Sometimes, extensive hemolysis causes anemia which may lead to death of the patient. PEGlyation or functionalization of the surfaces of NPs may provide a steric environment which prevents binding of NPs with opsonising proteins and thereby significantly increase circulation time of NPs in human body with high level of therapeutic index.

One of the primary steps in the treatment for bronchogenic carcinomas is its early detection; if possible at stage I. The available diagnostic methods (CT scanning, chest X-ray screening, etc.) may detect lung cancer only at stage IV which is too late for the survival of a patient. Let us imagine a developing technology where the first cancer cell may be detected and be removed before it is ever developed into a late-stage cancer. Let a broken component of a cell be removed/repaired/replaced with a miniature biological machine (nanorobot) or small-sized pumps of the order of biomolecules when implanted into the body by delivering lifesaving drugs precisely when and where needed. Emerging nanomedicine holds a promising area in adding colors to this imagination through synergistic action of nanotechnology, molecular therapeutics (gene therapy, small molecules therapy, antisense therapy, etc.), and conventional therapeutics (chemotherapy, radiotherapy, PDT, etc.). The ability of the nanoscale devices to simultaneously interact with the multiple proteins and nucleic acids at the molecular scale enables efficient signaling to differentiate between the behavior of the cells in their normal state and as they undergo malignant transformation. Such devices can directly recognize specific protein structures and structural domains or follow protein transport among different cellular compartments or any mutation in genetic or DNA structure as contrary to the normal healthy cells, due to the presence of adsorbed identifier molecules.

NPs used in biomedical field are broadly categorized under two headings:

- Organic nanoparticles
- Inorganic nanoparticles

Liposomes, solid lipid NPs, dendrimers, chitosan, virus, and polymeric NPs belong to the group of organic NPs. They are well established in the world of cancer imaging and treatment. But in the recent years, inorganic NPs have emerged as potential vector elements in cancer treatment. These include NPs of gold, silver, silica, magnetic particles, carbon particles, ceramic particles, and quantum dots. Major construction of this type of particle is a central elemental core that exhibits fluorescence, optical, magnetic, or electronic properties with a protective biocompatible, organic coating on the surface. Obviously, the outer organic coating protects the underlying core from degradation in physiologically harsh environment.

References

1. http://www.cancerresearchuk.org/cancer-info/cancerstats/world/cancer-worldwide-the-global-picture
2. http://www.medicinenet.com/lung_cancer/article.htm
3. http://www.cancerresearchuk.org/cancer-info/cancerstats/faqs/

4. Travis WD, Linder J, Mackay B (1996) Classification, histology, cytology, and electron microscopy. Pass HI, Mitchell JB, Johnson DH, Turrisi AT (eds) Lung cancer: principles and practice. Lippincott-Raven Publishers, Philadelphia, pp 361–395

5. Rani D, Somasundaram VH, Nair S, Koyakutty M (2012) Advances in cancer nanomedicine. J Indian Inst Sci 92:187–218

6. http://kkkmedicine.blogspot.in/2010/03/treatment-of-lung-cancer-by-radiation.html

7. Ahmad MB, Tay MY, Shameli K, Hussein MZ, Lim JJ (2011) Green synthesis and characterization of silver/chitosan/polyethylene glycol nanocomposites without any reducing agent. Int J Mol Sci 12:4872–4884

8. Sattler KD (2010) Handbook of nanophysics: nanomedicine and nanorobotics. CRC Press, Boca Raton

9. Jong WHD, Borm PJ (2008) Drug delivery and nanoparticles: applications and hazards. Int J Nanomed 3:133–149

10. Mohanraj VJ, Chen Y (2006) Nanoparticles—a review. Trop J Pharm Res 5:561–573

11. Calvo P, Remunan-Lopez C, Vila-Jato JL, Alonso MJ (1997) Chitosan and chitosan/ethylene oxide-propylene oxide block copolymer nanoparticles as novel carriers for proteins and vaccines. Pharm Res 14:1431–1436

12. Sivasankar M, Kumar BP (2010) Role of nanoparticles in drug delivery system. Int J Pharm Biomed Res 1:41–66

13. Bhattacharyya S, Kudgus R, Bhattacharya R, Mukherjee P (2011) Inorganic nanoparticles in cancer therapy. Pharm Res 28:237–259

14. Ng VWK, Berti R, Lesage F, Kakkar A (2013) Gold: a versatile tool for in vivo imaging. J Mater Chem B 1:9–25

15. Niidome T, Yamagata M, Okamoto Y, Akiyama Y, Takahashi H, Kawano T, Katayama Y, Niidome Y (2006) PEG-modified gold nanorods with a stealth character for in vivo applications. J Controlled Release 114:343–347

16. Kawano T, Yamagata M, Takahashi H, Niidome Y, Yamada S, Katayama Y, Niidome T (2006) Stabilizing of plasmid DNA in vivo by PEG-modified cationic gold nanoparticles and the gene expression assisted with electrical pulses. J Controlled Release 111:382–389

17. Goya GF, Grazu V, Ibarra MR (2008) Magnetic nanoparticles for cancer therapy. Curr Nanosci 4:1–16

18. Fuentes M, Mateo C, Guisan JM, Fernandez-Lafuente R (2005) Preparation of inert magnetic nano-particles for the directed immobilization of antibodies. Biosens Bioelectron 20:1380–1387

Chapter 2
Natural and Semisynthetic Nanoparticles in Lung Cancer Diagnosis and Therapy

Abstract Natural and semisynthetic nanoparticles have significantly attracted lung cancer therapy owing to their high biocompatibility and biodegradability. Most of them are suitable as carriers of drugs, genes, imaging probes, and various macromolecules. This chapter focuses on utilitarian properties with examples (both preclinical and clinical trials) of various natural and semisynthetic nanoparticles for the early-stage diagnosis and treatment of lung cancer.

Keywords Virus nanoparticle · Liposomes · Solid lipid nanoparticle · Chitosan nanoparticle · Polyherbal nanoparticle · Target-specific drug delivery · Gene therapy

2.1 Virus (Capsid) Nanoparticles

Viruses are infectious pathogens, ranging in size from 20 to 400 nm that encapsulate genome (nucleic acid) in a protein coat. Virus NPs are particles composed of virus capsid (shells) but devoid of genome. Hence, virus NPs are incapable of replication and non-infectious. Capsid is basically a protective coating that prevents virus NPs against extreme temperatures, range of pH, and various harsh chemicals. They are structurally symmetrical, polyvalent, and monodispersed. Also, they combine the advantages of non-immunogenity and biodegradability [1]. Interestingly, virus NPs may be covalently conjugated to variety of moieties ranging from drugs, targeting reagents, imaging probes to various inorganic NPs. Hence, these non-infectious forms are very effective at delivering therapeutic proteins or other vital materials into the target locations like cancer cells. Viruses also self-assemble naturally from coat protein monomers and encapsulate negatively charged nucleic acids. Hence, virus NPs may be exploited to trap artificial therapeutic nucleic acids and other polyanions [2]. Mammalian viruses pose a significant threat of infection and thus are deliberately avoided. This is a major reason to consider plant viruses as an alternative for biomedical applications. Plant viruses do not infect mammalian cells. Also, they have an extremely low probability of genetic recombination with animal viruses and

© The Author(s) 2015
A. Bandyopadhyay et al., *Nanoparticles in Lung Cancer Therapy - Recent Trends*,
SpringerBriefs in Molecular Science, DOI 10.1007/978-81-322-2175-3_2

no inherent capability of targeting any biological entity until and unless chemically modified. Commonly employed virus vectors include cowpea mosaic virus (CPMV), tobacco mosaic virus (TMV), cowpea chlorotic mottle virus, canine parvovirus, and bacteriophages such as Qβ and MS$_2$. Virus NPs are advantageous over syntheti-cally designed NPs with respect to precise dimensions, possible evasion of immune system, biocompatibility, and biodegradability [3]. CPMVs capsid proteins are most widely studied viral vectors. They exhibit icosahedrons structure with a spherical aver-age size of 28.4 nm, formed by 60 identical subunits and are stable up to a tempera-ture of 60 °C and in the pH range of 3.5–9 [4]. CPMVs capsid contains no cysteine group on the exterior surface; thus, they lack any naturally occurring thiol groups. Since CPMVs capsid consists of 60 identical subunits, each amino acid modification may be introduced into 60 positions on the viral surface. Thus, a single added cysteine (HO$_2$CCH(NH$_2$)CH$_2$SH) produces 60 reactive thiols on the capsid, while adding two cysteines produce 120 reactive thiols [4]. These thiol groups on the exterior surface of CPMVs capsid encourage attachment of thiol reactive gold NPs via strong sul-fur–gold covalent bonds and help in delivering highly contrast agents to the affected zones for sensitive MRI detection (say in case of imaging of deeply embedded lung cancer cells). In another study, Robertson et al. [5] mentioned that bacteriophage T4 NPs could be converted into fluorescent imaging probes via conjugation between amine groups in lysine on T4 NPs surface and various organic dyes like FITC, Cy3 NHS ester, and Alexa Fluor 546 NHS ester dyes. They successfully utilized these sens-ing probes for in vitro cellular imaging and flow cytometry. They clearly demonstrated cellular uptake of dye-T4 NPs in the lung cancer cell line A549 (human lung adeno-carcinoma) under confocal microscope at different incubation times (Fig. 2.1a). DAPI staining (blue) indicated the nuclei of cells and the small yellow spots distributed

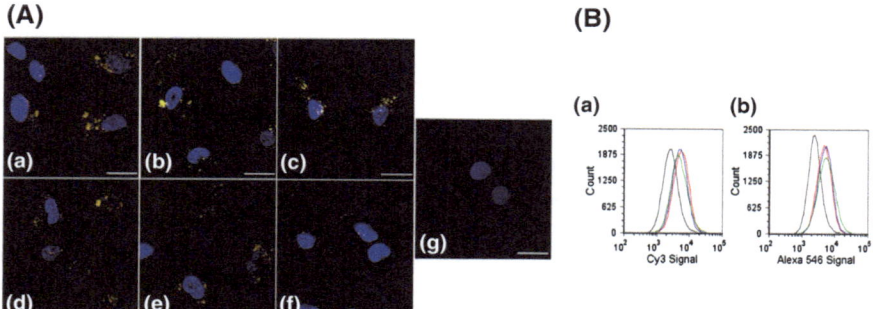

Fig. 2.1 **A** Confocal microscopy images of A549 cells after uptake of dye-T4 NPs at different time points. Uptake of Alexa 546-T4 NPs (2282 D/V): **a** 4 h after incubation, **b** 8 h, and **c** 24 h. Uptake of Cy3-T4 NPs (786 D/V) after incubation at **d** 4 h, **e** 8 h, **f** 24 h, and **g** untreated cells, negative control. Imaging was performed under 60× magnification. The scale bar is 10 μm. **B** Flow cytometry histograms of A549 cells treated **a** Cy3-T4 NP (786 D/V) and **b** Alexa-T4 NP (2282 D/V) at different time points after incubation, 4 h (*red*), 8 h (*blue*), and 24 h (*green*); untreated cells after 24 h incubation are used as negative control (*black line*). Reprinted (adapted) with permission from Robertson et al. [5]. Copyright (2011) American Chemical Society

throughout the cells were actually the dye-T4 NPs. Even they obtained interesting results of cellular uptake of dye-T4 NPs in flow cytometry experiments by measuring the difference in fluorescence signal between the treated and untreated cells (Fig. 2.1b). Their preclinical study showed that dye-T4 NPs stayed in the A549 cells for at least 72 h and enabled cell tracking owing to their fluorescence property.

2.2 Liposomes and Solid Lipid Nanoparticles

Conventional liposomes are self-assembling concentric lipid bilayered vesicles (highly hydrophobic) surrounded by a non-toxic amphiphilic phospholipid membrane, engulfing an aqueous core (Fig. 2.2) [6, 7]. They are the most studied non-viral carriers for delivery of drugs and other macromolecules to various targeting cells. Owing to the high hydrophobicity of lipid bilayer, they may be loaded with a variety of molecules including drug molecules, proteins, nucleic acids, and even plasmids. In fact, encapsulation of fluorometrically detectable dyes, enzymes, and electro-active compounds within liposomes and their successful delivery to damaged cells by EPR effect dramatically help in early detection of any form of cancer cells. Lipophilic drugs get tightly incorporated into the phospholipid bilayer of the liposomes, thereby enhance the hydrophobic drug solubility in the bloodstream. Even sometimes, specific antibodies or proteins (specific to a certain receptor or cancerous antigen) may be conjugated to their outer surface to improve targeting to specific cells. Currently there are only two clinically tried food and drug administration (FDA) approved liposomal formulations, DOXIL (a liposomal doxorubicin administration for ovarian cancer) and Marqibo (a liposomal vincristine sulfate administration for lymphoblastic leukemia) [8]. Lipusu (a liposomal PTX formulation) is also available in the market for clinical trials. However, examples of liposomes formulations as potential therapeutic systems for NSCLCs are limited. Cisplatin drug which is highly effective against NSCLC cells are often found to be nephrotoxic. Boulikas and his team developed lipoplatin (cisplatin conjugated to liposomes) which were found to significantly reduce nephrotoxicity in rat tumor

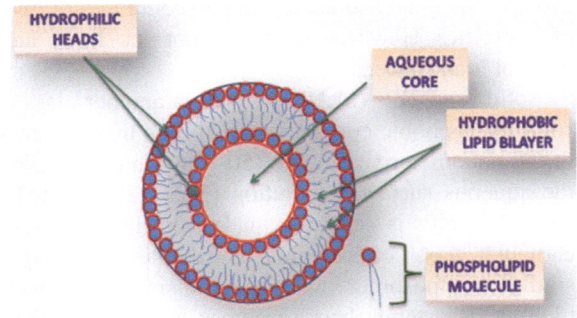

Fig. 2.2 Hydrophilic and hydrophobic layers in liposomes

models [9]. Hence, in future, lipoplatin may be effectively used for the treatment of lung cancer. In fact, a recent report claims that lipoplatin may successfully complete phase III clinical trial testings by 2014 [10]. Even liposomal delivery of PTX to NSCLC cells for phase I clinical trials demonstrated significant improvement in therapy. Interestingly, in a preclinical trial, a liposome PTX formulation successfully targeted lung cancer cells and reduced incidence of drug resistance. At the present scenario, researchers are trying to develop more liposomal formulations for treatment of metastatic cancers too. Liposomes may also be used in immunotherapy to deliver cancer vaccines. Tumor-associated antigenic (TAAs) stimuli may be encapsulated in the aqueous core or embedded in the bilayer or attached to the outer surface of liposomes [11]. In terms of clinical trials, Sangha and North showed that therapeutic vaccine Biomira Liposomal Protein 25 (BLP25 or Stimuvax® which comprises of conventional liposomes, lipid A adjuvant and tumor-associated antigen-pulmitolyted MUC1 peptide) provided satisfactory results in the treatment of advanced stage NSCLCs (phase III) and survival of patients [12]. One of the major problems associated with conventional liposomal delivery systems is short systemic half cycle; they are rapidly cleared by RES. Stealth liposomes or long circulating liposomes which are coated with steric and hydrophilic molecules (PEG, glycolipids etc.), enable them to encapsulate hydrophilic (mitoxantrone etc.) as well as hydrophobic (doxorubicin etc.) chemotherapeutic drugs into the aqueous core and consequently enhance the drug circulation time in the bloodstream without any loss of drug molecules. Stealth coating creates a neutral layer on the surface of liposomes which prevent any undesirable interaction with biological proteins (i.e., prevents opsonization). In fact, stealth-coated liposomes are widely used in any liposomal formulations to prevent their undesirable clearance from the human body. Altin and his coworkers showed that APCs-based dendritic cells (DCs) targeting stealth-coated liposomes encapsulated with OVA ± LPS or IFN-γ when vaccinated to mice models successfully treated lung metastases by OVA expressing B16 melanoma cells [13]. However, many recent studies report that unfortunately despite the use of stealth strategies, often liposomes are filtered out through the hepatic portal system. Hence, stealth liposomes are further functionalized with desirable targeting ligands (antibodies or other moieties directed to specific antigens, receptors, or other targets on the pulmonary endothelium) to ensure more accumulation of the NPs in lung cancer cells. Cationic liposomes (composed of charged heads and hydrophobic carbon skeletons) and archeosomes (composed of glycerolipids) are other variations of liposomes. However, they have not yet shown any successful results in clinical trials of lung cancer.

Solid lipid NPs (SLNPs) which is another class of lipid NPs are made up of a solid hydrophobic core surrounded by a monolayer of phospholipid coating, often in the nanometer size range (50–1,000 nm) and are well dispersed in water or in the aqueous surfactant solution (Fig. 2.3) [14, 15]. The solid hydrophobic lipid core of SLNPs is generally made from lipid molecules of triglycerides, beeswax, carnauba wax, cetyl alcohol, and cholesterol [16]. SLNPs are structurally quite similar to liposomes and nanoemulsions. However, SLNPs are more stable in biological systems than liposomes and nanoemulsions owing to the presence of

Fig. 2.3 Layers in SLNPs

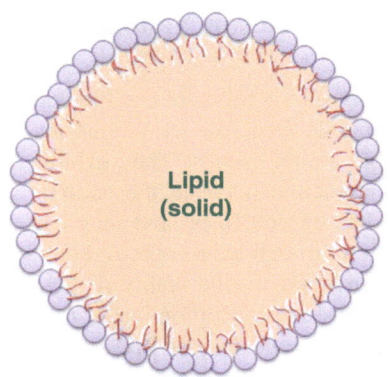

rigid hydrophobic core. SLNPs are often referred to as "zero-dimensional" nano-materials owing to their all dimensions in the nanometer range [17]. Due to solid hydrophobic environment of the core, SLNPs are often used as colloidal carrier of hydrophobic chemotherapeutic drugs and genes for long-term circulation in the bloodstream. Mutation and subsequent loss in the ability of the p53 tumor suppressor gene to induce growth arrest results in apoptosis (programmed cell death) which is the primary reason behind endobronchial cancer [17–19]. Thus, the disease could be cured by the transfer of wild type p53 gene to the defective and deficient tumors present in the lower respiratory airways. Cationic SLNPs are amphiphilic in nature owing to the presence of one or two hydrophobic fatty acid side chain and a hydrophilic amino group with a linker. The hydrophilic amino group interacts spontaneously with the negatively charged DNA plasmids to form a stable complex and promotes effective gene transfer to cells. Thus, p53/cationic SLNPs complex may be administered into the affected areas either by intra-tracheal instillation method (genes are delivered directly through an airway catheter) or by aerosol inhalation [19]. Repeated administration may be recommended as it ensures removal of superficial apoptotic cells and enables direct access of the complexes to deeper layers of epithelial tissues lining the endotracheal tube. Even Choi et al. [20] quite efficiently transfected p53 tumor suppression genes to null H1299 lung cancer cells using SLNPs delivery shuttles. Drugs or genes are generally loaded into SLNPs at low temperature and unloaded at higher temperatures. Thus, by hyperthermia effect of tumors microenvironment, SLNPs may successfully unload therapeutic macromolecules in specific areas [21]. Interestingly, SLNPs may be used to load chemotherapeutic drugs, tumor suppressor genes, and imaging probes simultaneously. In a recent study, researchers successfully loaded SLNPs with Bcl-2-siRNA, PTX, and CdSe/ZnS quantum dots. SLNPs provide some significant advantages over liposomes which include less drug leakage in the bloodstream, prevention of drug hydrolysis, prolonged drug circulation, and controlled drug delivery due to solid lipid matrix which immobilize the drug molecules. Thus, in the near future SLNPs may prove to be highly promising anticancer kits especially for the detection and eradication of impenetrable lung cancer cells.

2.3 Chitosan Nanoparticles

Chitosan is a biocompatible and biodegradable cationic and nitrogenous polysaccharide (obtained by partial deacetylation of chitin) bearing a number of ionisable amino groups. They exhibit mucoadhesive nature due to strong electrostatic interaction between the cationic chitosan molecules and the negatively charged sites on cell lines, and they can promote encapsulated macromolecules permeation through well-organized epithelial cells. Thus, chitosan molecules are lucrative candidates in the realm of therapeutics delivery. Nucleic acids being negative in nature are mostly delivered to the lung cancer cells using chitosan NPs. Another important feature of chitosan is that they are readily degraded by lysozyme which is the most abundant enzyme in the lung cells and thus cleared from the human body after gene transfection or drug delivery. Lv et al. [22] in their novel work loaded a water insoluble mitotic cell cycle arresting drug PTX into N-((2-hydroxy-3-trimethylammonium)propyl) chitosan chloride (HTCC) NPs. They treated the Lewis lung cancer (LLC) cells in xenografted mouse models with various formulations of PTX (Taxol, CS-NP: PTX, and HTCC-NP: PTX) and found that the tumor volume decreased with time after treatment due to successful accumulation of HTCC-NPs: PTX in tumor locations by EPR effect. Figure 2.4 shows that HTCC-NP: PTX exhibited maximum tumor inhibition efficacies.

In another work, Galbeiti and his coworkers prepared a conjugate of folic acid bound chitosan on the surface of polyvinyl alcohol microcapsules (MC-Chi-FA) [23]. They loaded camptothecin (CPT) drug (frequently used in the treatment of SCLC recurring diseases) into the MC-Chi-FA conjugate and were successful to target the tumors of epithelial origin (e.g., HeLa cells overexpressing the folate receptor) with the conjugate of MC-Chi-FA/CPT. Okamoto et al. [24] synthesized gene powders of a luciferase expression plasmid driven by the cytomegalovirus promoter (pCMV-Luc) in low molecular weight chitosan nanoparticle vectors using carbon dioxide as supercritical fluid. They successfully transfected the pCMV–Luc genes to the lung cancer cells, and the chitosan-pDNA powders exhibited very high pulmonary luciferase activity. Ventura and his companions designed $2'$, $2'$-difluorodeoxycytidine (GEM—a deoxycytidine analog used in the treatment of NSCLC) loaded chitosan microspheres by the spray drying technique using different amounts of polyanions dextran sulfate (DS) [25]. These GEM-Chi-DS microspheres exhibited promising antitumoral efficacy in vitro on human lung cancer cells (A549). Boca et al. [26] synthesized chitosan-coated silver nanotriangles (Chit-AgNTs) which behaved as photothermal agents against a line of human NSCLC cells (NCI-H460) owing to the strong resonances of AgNPs in the near infrared (NIR) region. During cytotoxicity assays, chitosan-coated triangular silver NPs were efficiently taken by the cancer cells and exhibited low cytotoxicity on normal embryonoic cells. Nafee and his groups in another work embellished chitosan-modified PLGA NPs with antisense oligonucleotide, $2'$-O-Methyl-RNA (OMR—a potential telomerase inhibitor which prevents telomere shortening during cell proliferation in NSCLC tissues) and successfully transfected OMR to A549 cells and Calu-3 cells [27]. Often various

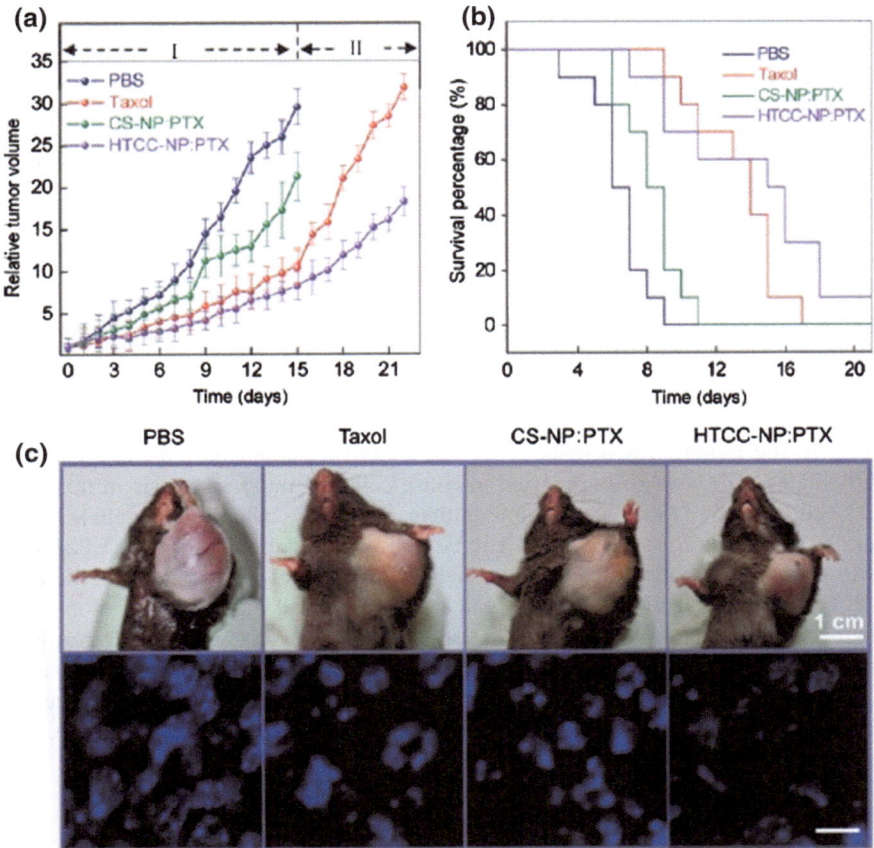

Fig. 2.4 Antitumor efficacies of different PTX formulations in the subcutaneous mouse model of LLC. **a** Tumor volumes of mice during (*I*) and after (*II*) treatment with different PTX formulations. **b** Survival of mice in different treatment groups. **c** Tumor images and the corresponding cell nuclear of different groups after treatment. *Scale bar* 10 μm. Reprinted (adapted) with permission from Lv et al. [22]. Copyright (2011) American Chemical Society

polymer NPs are modified with chitosan for oral delivery of chemotherapeutic drugs and genes to deeply embedded lung cancer cells. Earlier oral delivery of anticancer agents for the treatment of lung cancer was not practiced because most of the therapeutic molecules accumulated in liver or spleen, caused tremendous discomfort in patients and finally cleared by RES. Jiang et al. [28] introduced chitosan-modified PCL at the surface for oral delivery of chemotherapeutic drugs to lung cancer cells and showed that high mucoadhesive properties of surface chitosan enabled selective interaction of anticancer drugs with mucin which is over expressed in any cancer cells. They studied that chitosan modification introduced positive charges on poly (lactic acid) and poly (lactic-co-glycolic) acid NPs which facilitated rapid cellular uptake and high cytotoxicity against lung cancer cells. Mehrotra et al. [29] showed that lomustine (an antineoplastic agent) loaded chitosan NPs exhibited high cytotoxicity against lung cancer cells L132 by very controlled release of lomustine. Thus, chitosan-modified NPs may be designed as potential delivery shuttles for drugs, genes etc. Chitosan-modified NPs also improve sustained delivery of therapeutic molecules.

2.4 Polyherbal Nanoparticles

Often herbal drugs are preferred for the treatment of diseases owing to the absence of cytotoxicity and subsequent side effects in human. As tumor architecture causes NPs to preferentially accumulate at the tumor site, their implication as a drug delivery vector has raised attention in delivering and localizing greater amount of drug to the target sites. Ethanolic extract of Polygala senega (EP) and Zingiber officinalis (EZ) have been reported to cause cell death and apoptosis in lung cancer cell line A549 (adenocarcinomic human alveolar basal epithelial cells). Jadhav in his work mentioned that when EP, EZ, and combination of both (CEPZ) were encapsulated in biodegradable polymer NPs such as poly(lactic-co-glycollic acid) (PLGA) enhanced bioavailability, and cellular uptake was observed along with induced apoptosis of A549 cells, decreased expression of survivin, PCNA mRNA and increased expression of caspase-3 and p53 mRNAs [30]. However, research and clinical trials for treatment of cancer using polyherbal NPs is very rare. We believe that large scale synthesis of polyherbal NPs is not possible. Hence, they have not been commercialized.

2.5 Other Natural Nanoparticles

Nanoemulsions (NEs) are stable colloidal dispersions of oil in water. NEs are generally stabilized by biocompatible surface active agents. NEs vary in size from 20 to 200 nm. Often NEs are used in immunotherapy owing to their long circulation time in human bloodstream and rapid uptake by APCs [11]. Their use in lung cancer vaccination is not yet reported because they fail to target deeply embedded lung cancer cells. However, more extensive research may render NEs successful in lung cancer therapy.

Some protein-based NPs like those of gelatin, dextran, albumin etc., are widely studied either alone or in conjunction with biodegradable polymers as delivery systems. Wiley et al. [31] synthesized protein cage NPs (PCN) from small heat shock proteins (sHsp 16.5) of hyperthermophilic archaeon Methanococcus jannaschii and successfully enhanced immune responses by increasing formation of inducible bronchus-associated lymphoid tissue (iBALT) on exposure of PCNs to pulmonary cell lines. This reduced pulmonary viral infection during treatment of lung cancer which may otherwise worsen the situation. Abraxane (a FDA approved albumin based NPs carrying PTX) is often currently used in first line treatment of advanced and metastatic NSCLCs in combination with carboplatin where patients cannot be operated or exposed to radiation due to various internal complexities [32]. However, natural polymers such as proteins or polysaccharides are avoided in delivering drugs and genes owing to their variance in purity and risk of denaturation of the embedded drug molecules on cross-linking which is often required to support the prepared structures. Natural NPs also lack reproducibility and hence not suitable for commercialization.

References

1. Cormode DP, Jarzyna PA, Mulder WJM, Fayad ZA (2010) Modified natural nanoparticles as contrast agents for medical imaging. Adv Drug Deliv Rev 62:329–338
2. Pokorski JK, Steinmetz NF (2011) The art of engineering viral nanoparticles. Mol Pharm 8:29–43
3. Rani D, Somasundaram VH, Nair S, Koyakutty M (2012) Advances in cancer nanomedicine. J Indian Inst Sci 92:187–218
4. Blum AS, Soto CM, Wilson CD, Cole JD, Kim M, Gnade B, Chatterji A, Ochoa WF, Lin T, Johnson JE, Ratna BR (2004) Cowpea mosaic virus as a scaffold for 3-D patterning of gold nanoparticles. Nano Lett 4:867–870
5. Robertson KL, Soto CM, Archer MJ, Odoemene O, Liu JL (2011) Engineered T4 viral nanoparticles for cellular imaging and flow cytometry. Bioconjug Chem 22:595–604
6. Caruso G, Caffo M, Alafaci C, Raudino G, Cafarella D, Lucerna S, Salpietro FM, Tomasello F (2011) Could nanoparticle systems have a role in the treatment of cerebral gliomas? Nanomed Nanotech Biol Med 7:744–752
7. Wang X, Yang L, Chen Z, Shin DM (2008) Application of nanotechnology in cancer therapy and imaging. CA Cancer J Clin 58:97–110
8. Babu A, Templeton AK, Munshi A, Ramesh R (2013) Nanoparticle-based drug delivery for therapy of lung cancer: progress and challenges. J Nano Mat 2013:11
9. Boulikas T (2004) Low toxicity and anticancer activity of a novel liposomal cisplatin (Lipoplatin) in mouse xenografts. Oncol Rep 12:3–12
10. Liposomal cisplatin (Nanoplatin) for advanced non-small cell lung cancer—first line (2012) http://www.hsc.nihr.ac.uk/topics/liposomal-cisplatin-nanoplatin-for-advanced-non-sm/
11. Krishnamachari Y, Geary S, Lemke C, Salem A (2011) Nanoparticle delivery systems in cancer vaccines. Pharm Res 28:215–236
12. Sangha R, North S (2007) L-BLP25: a MUC1-targeted peptide vaccine therapy in prostate cancer. Expert Opin Biol Theo 7:1723–1730
13. van Broekhoven CL, Parish CR, Demangel C, Britton WJ, Altin JG (2004) Targeting dendritic cells with antigen-containing liposomes: a highly effective procedure for induction of antitumor immunity and for tumor immunotherapy. Cancer Res 64:4357–4365
14. Ekambaram P, Sathali AAH, Priyanka K (2012) Solid lipid nanoparticles: a review. Sci Revs Chem Commun 2:80–102
15. Shah CV, Shah V, Upadhyay U (2011) Solid lipid nanoparticles: a review. CPR 1:351–368
16. Barthelemy P, Laforet JP, Farah N, Joachim J (1999) Compritol® 888 ATO: an innovative hot-melt coating agent for prolonged-release drug formulations. Eur J Pharm Biopharm 47:87–90
17. Mathur V, Satrawala Y, Rajput MS, Kumar P, Shrivastava P, Vishvkarma A (2010) Solid lipid nanoparticles in cancer therapy. Int J Drug Deliv 2:192–199
18. Choi SH, Jin SE, Lee MK, Lim SJ, Park JS, Kim BG, Ahn WS, Kim CK (2008) Novel cationic solid lipid nanoparticles enhanced p53 gene transfer to lung cancer cells. Eur J Pharm Biopharm 68:545–554

19. Zou Y, Zong G, Ling Y-H, Hao MM, Lozano G, Hong WK, Perez-Soler R (1998) Effective treatment of early endobronchial cancer with regional administration of liposome-p53 complexes. J Natl Cancer Inst 90:1130–1137

20. Choi SH, Jin SE, Lee MK, Lim SJ, Park JS, Kim BG, Ahn WS, Kim CK (2008) Novel cationic solid lipid nanoparticles enhanced p53 gene transfer to lung cancer cells. Eur J Pharm Biopharm 68:545–554

21. Thakor AS, Gambhir SS (2013) Nanooncology: the future of cancer diagnosis and therapy. CA Cancer J Clin 63:395–418

22. Lv PP, Wei W, Yue H, Yang TY, Wang LY, Ma GH (2011) Porous quaternized chitosan nanoparticles containing paclitaxel nanocrystals improved therapeutic efficacy in non-small-cell lung cancer after oral administration. Biomacromolecules 12:4230–4239

23. Galbiati A, Tabolacci C, Morozzo Della Rocca B, Mattioli P, Mattioli S, Beninati G, Paradossi G, Desideri A (2011) Targeting tumor cells through chitosan-folate modified microcapsules loaded with camptothecin. Bioconjug Chem 22:1066–1072

24. Okamoto H, Nishida S, Todo H, Sakakura Y, Iida K, Danjo K (2003) Pulmonary gene delivery by chitosan–pDNA complex powder prepared by a supercritical carbon dioxide process. J Pharm Sci 92:371–380

25. Ventura CA, Cannava C, Stancanelli R, Paolino D, Cosco D, La Mantia A, Pignatello R, Tommasini S (2011) Gemcitabine-loaded chitosan microspheres. Characterization and biological in vitro evaluation. Biomed Microdevices 13:799–807

26. Boca SC, Potara M, Gabudean A-M, Juhem A, Baldeck PL, Astilean S (2011) Chitosan-coated triangular silver nanoparticles as a novel class of biocompatible, highly effective photothermal transducers for in vitro cancer cell therapy. Cancer Lett 311:131–140

27. Nafee N, Schneider M, Friebel K, Dong M, Schaefer UF, Murdter TE, Lehr CM (2012) Treatment of lung cancer via telomerase inhibition: self-assembled nanoplexes versus polymeric nanoparticles as vectors for 2′-O-Methyl-RNA. Eur J Pharm Biopharm 80:478–489

28. Jiang L, Li X, Liu L, Zhang Q (2013) Thiolated chitosan-modified PLA-PCL-TPGS nanoparticles for oral chemotherapy of lung cancer. Nanoscale Res Lett 8:66

29. Mehrotra A, Nagarwal RC, Pandit JK (2011) Lomustine loaded chitosan nanoparticles: characterization and in-vitro cytotoxicity on human lung cancer cell line L132. Chem Pharm Bull (Tokyo) 59:315–320

30. Jadhav UG (2012) Fabrication of polyherbal nanoparticles—a targetted herbal drug delivery for lung cancer. Pharm Anal Acta; Pharmaceutica 3

31. Wiley JA, Richert LE, Swain SD, Harmsen A, Barnard DL, Randall TD, Jutila M, Douglas T, Broomell C, Young M, Harmsen A (2009) Inducible bronchus-associated lymphoid tissue elicited by a protein cage nanoparticle enhances protection in mice against diverse respiratory viruses. PLoS ONE 4:e7142

32. Cho K, Wang X, Nie S, Chen Z, Shin DM (2008) Therapeutic nanoparticles for drug delivery in cancer. Clin Cancer Res 14:1310–1316

Chapter 3
Synthetic (Organic) Nanoparticles Induced Lung Cancer Diagnosis and Therapy

Abstract Natural or semisynthetic nanoparticles often degrade too fast and are cleared by the hepatic portal system before they effectively deliver therapeutic macromolecules especially to lung cancer cells. In an attempt to prolong shelf life of nanovectors in the human bloodstream for enhanced therapeutic efficacy of anticancer agents against cancer cells, nanoparticles are prepared synthetically. This chapter focuses on different synthetically prepared organic nanomaterials which have shown successful preclinical results in the treatment of lung cancer.

Keywords Polymer nanoparticle · Micelles · Dendrimer · Carbon nanoparticle · Passive targeted therapy

3.1 Polymer Nanoparticles (Synthetic)

Polymer NPs (PNPs) are colloidal particles (50–300 nm), prepared from biocompatible and biodegradable synthetic polymers like poly(lactic acid) (PLA), poly(glycolic acid) (PGA), poly(lactic-co-glycolic) acid (PLGA), poly(L-glutamic acid), poly(ε-caprolactone) (PCL), poly(amino acids), poly(ethylene glycol) (PEG), poly(alkyl cyanoacrylates), N-(2-hydroxypropyl) methacrylamide copolymer, and poly(styrene maleic anhydride) copolymer. Broadly, PNPs used in drug/gene delivery are of two types: polymer nanospheres and polymer nanocapsules. Nanospheres are spherical and solid NPs where macromolecules are either adsorbed or covalently attached to the surface. Nanocapsules are polymeric vesicles consisting of a core of water or oil and surrounded by a polymeric shell where macromolecules are encapsulated inside the core [1]. PNPs exhibit very high chemical stability, are easy to fabricate in large quantities, may be administered both orally and intravenously with significant efficiency, and enable specific targeting to cancer cells by strong EPR effect. PNPs significantly enhance chemo- and radiotherapy efficacies of various anticancer agents. Jung et al. [2] reported that PEG-modified PLA NPs loaded with taxanes improved therapeutic index of

© The Author(s) 2015
A. Bandyopadhyay et al., *Nanoparticles in Lung Cancer Therapy - Recent Trends*,
SpringerBriefs in Molecular Science, DOI 10.1007/978-81-322-2175-3_3

chemoradiotherapy (which involve simultaneous administration of chemotherapeutic drugs and radiation to patients with lung cancer) both in vitro and in vivo of A549 cells in xenografted models. PLGA-based NPs being US FDA approved are often used to deliver antineoplastic drugs, tumor suppressor genes, DNA, and other biomacromolecules to A549 lung cancer cells. However, in vivo studies revealed that although PLGA is an excellent drug/gene carrier with high target-specific efficiency toward cancer cells, in complicated human circulatory systems, they are prone to rapid clearance out of the body. Hence, often, PEG coating on PLGA NPs is done to prolong the blood circulation time of the NPs. Sengupta et al. [3] synthesized a drug delivery shuttle composed of a core of PLGA NPs enveloped inside PEGlylated-phospholipid block copolymer. They simply trapped antiangiogenesis agent combretastatin-A4 within the outer lipid envelope and conjugated cytotoxic agent DOX to the PLGA NPs core. When this antineoplastic agents-loaded delivery shuttles reached tumors, at first the outer protective layer of phospholipid disrupted and released the antiangiogenesis agents which caused rapid vascular shutdown owing to the collapsing of cytoskeletal structures. Simultaneously, PLGA NPs got trapped inside the tumors and slowly released the cytotoxic agents to kill the neoplastic cells by inducing apoptosis. This synergestic effect of anticancer agents was successfully developed by Sengupta et al. to treat NSCLC-type bronchogenic cancer in mice. In another in vitro study, Benfer and Kissel used cationic PLGA [PLGA grafted to 3-(diethylamino) propylamine (DEAPA) modified PVA) to deliver negatively charged siRNA (small interfering RNA) as gene-silencing agent to inhibit protein synthesis essential for tumor growth of H1299–EGFP cells (lung cancer cells expressing green fluorescent protein] [4]. Poly(ethylene imine) (PEI) is an interesting class of organic polymer with high density of amino groups that can be protonated and thus extensively complexes with various negatively charged nucleic acid. PEI NPs are often utilized in delivery systems to transfect cancer cell lines with labeling nucleic acids. However, PEI is not biodegradable and improvement in their transfection efficiency by increasing molecular weight is often accompanied by cytotoxicity [5]. PEI used in gene transfection is often modified with biocompatible polymers like PEG, PLA, poly(caprolactone) or Pluoronics. Zhao et al. [6] mentioned that heparin-conjugated PEI system are biocompatible and could be used to deliver pIL therapeutic gene in inhibition of lung metastasis of B16–F10 malignant melanoma in murine model. In another work, Jere et al. grafted chitosan (a cationic polysaccharide with low cytotoxicity and low transfection efficiency) onto poly(ethylene imine) (a cationic polymer with high cytoxity and high transfection efficiency) for efficient delivery of Akt1 siRNA to A549 lung cancer cell line, silenced oncoproteins, and successfully inhibited tumor cell proliferation and tumorogenesis [7].

Often, PNPs are made up of di-/triblock copolymers of various biocompatible polymers as mentioned above. In an aqueous solution, amphiphilic homopolymers or diblock copolymers spontaneously assemble to form supramolecular morphologies like spherical micelles, wormlike micelles, and vesicles (polymersomes) [8–10]. Polymer micelles were developed based on the concept of aggregation of hydrophobic and hrdrophilic segments in liposomes and SLNPs. Most of the polymeric

NPs are micellar particles of diameter less than 1 μm. Micellar particles offer many advantages like low viscosity, small aggregate size, easy synthesis, and longer shelf life with prolonged circulation ability in bloodstream. In addition, they can permeate through small pores, reduce toxic side effects, and enable targeting. The degradation rate and the drug/gene/imaging probe release rate may be controlled in a sustained manner by clever manipulation of the ratio of the components of block copolymers and the hydrophilicity/hydrophobicity of the components.

3.1.1 Spherical Polymer Micelles

Polymeric spherical micelles are spherical assembly of hydrophilic and hydrophobic segments. Generally, the hydrophobic segments form the inner core and the hydrophilic segments the outer core owing to the large solubility difference of the two segments, as shown in the Fig. 3.1. Self-assembling is basically a thermodynamic process. Polymer micelles are thermodynamically stable aggregates. It is well known that the relative thermodynamic stability of polymer micelles is inversely proportional to the critical micelle concentration (CMC) of the same. This factor in turn is controlled by the length of the hydrophobic segment [11]. An increase in length of the hydrophobic block alone significantly reduces the CMC of the unimer construct, i.e., increases the thermodynamic stability and vice versa. Also, if the drug used is hydrophobic in nature, then the drug–hydrophobic core interaction further reduces the CMC and enhances the stability of the micelles [12]. So they are often used to load hydrophobic tumor-targeting drugs like doxorubicin

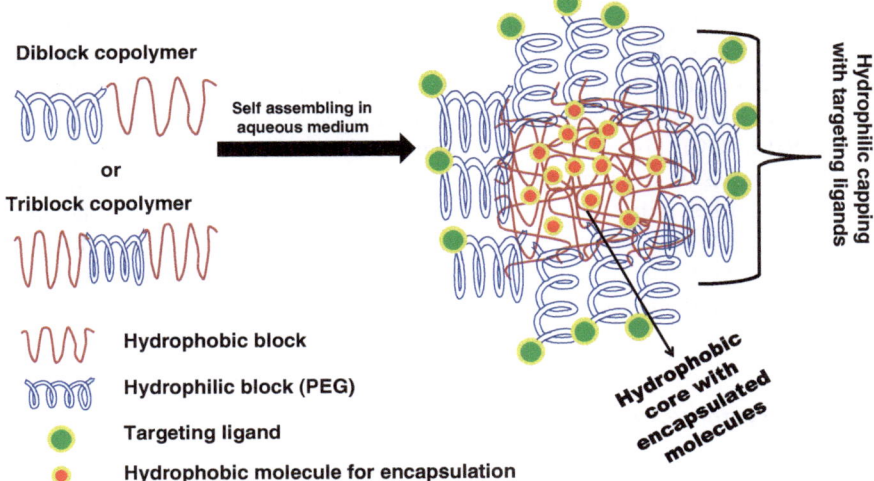

Fig. 3.1 A spherical polymeric micelle of block copolymer with targeting ligands

(DOX) either through physical encapsulation into the core or by chemical conjugation to the hydrophobic segments prior to micelle formation. Stealth coating and smaller size (50 nm–5 μm) of polymer micelles mask them from recognition by the RES and thereby prolong their circulation in the bloodstream. Some of the known examples such as poly(ethylene oxide)-block-poly(propylene oxide)-block-poly(ethylene oxide) or Pluronics®, poly(ethyl imine)-block-poly(lactic-co-glycolicacid), and poly(ethylene oxide)-block-poly(amino acid) have been successfully used as drug vehicles in lung tumor targeting. Poly(ethylene oxide) is the most commonly used hydrophilic block due to its high biocompatibility and polylactide or polycaprolactone on the other hand as the hydrophobic block due to easy hydrolytic degradation of the latter which is good for drug release mechanism [13].

3.1.2 Worm Like Polymer Micelles

An interesting class of supramolecular assemblies is cylindrical wormlike micelles. Unlike conventional spherical micelles, they are mostly micron in size, yet they can worm easily and are unhindered through small pores within our body owing to their small cross section and high flexibility. Targeted worms bind with high affinity to the surface of the targeted cells bearing receptor sites and thereby get internalized. On internalization, they deliver large amounts of drugs at a time to the affected locations and provide an instant effective healing effect as contrary to the spherical micelles. Actually, wormlike micelles have 1.5 times volume to surface area ratio than spherical micelles as derived from the geometric considerations, which results in very high drug loading inside the core of wormlike micelles. Such morphologies are predominantly possible if the weight fraction of the hydrophilic block is reduced below ~50 % of the total molecular weight of the copolymer [10]. One of the most commonly used block copolymer is poly(ethylene oxide)-block-PCL, which notably forms the worm configuration [10]. Other reported block copolymers suitable for generation of worm morphology are PEG-block-poly(ethyl ethylene), designated as OE and PEG-block-PCL, designated as OCL [14]. A study on successful internalization of targeted wormlike micelles inside the lung epithelial cells was reported by Discher et al. [14]. They even further utilized these filamentous carriers for loading chemotherapeutic agents (PTX) and released the drug selectively in the lung tumors. They also showed that wormlike polymer micelles exhibited significantly less cytotoxicity and greater potency in delivering TAX to human lung carcinoma A549 cells.

3.1.3 Polymersomes

Polymersomes constitute a class of artificial polymeric vesicles (i.e., engulf an aqueous core), composed of hydrophobic–hydrophilic diblock copolymers [15].

They vary in the size range from 50 nm to 5 μm. The polymeric core may encapsulate and protect sensitive molecules like drugs, enzymes, proteins, peptides DNA, and RNA molecules. Polymersomes are structurally similar to liposomes but vary in compositions. Unlike liposomes, they have polymeric bilayers. Thus, they are more mechanically stable and flexible than liposomes. They exhibit greater storage capabilities owing to their large hydrophobic core which follows from the large amphiphilic polymer membrane and prolonged circulation in the bloodstream [16]. Their assembling nature is, however, similar to viral capsids. The surfactant characteristics of the polymeric membranes (hydrophilic–hydrophobic block) results in spontaneous aggregation of the molecules in water into bilayers; the hydrophilic parts line the outer surface of the bilayer and fully shield the hydrophobic parts from the aqueous environment. There is a general rule that if the ratio of hydrophilic parts to the total copolymer mass is less than ~35 %, then it favors formation of membranous structure [17]. They can encapsulate both hydrophilic drug in their hydrophilic core and hydrophobic drug molecules within their thick lamellar membranes [18]. Waterhouse et al. used DOX-loaded poly(g-benzyl L-glutamate)-block-hyaluronan (PBLG-b-HYA)-based polymersomes (Poly-DOX) in the treatment of lung cancer. Free DOX when circulated in the human body tend to accumulate in the heart and causes cardiac toxicity. This problem is greatly alleviated by using polymersome carriers [19]. In many studies, both DOX and PTX have been loaded into the same polymeric vesicles of PEG-block-PLA and PEG-block-poly(caprolactone) to provide synergistic anticancer therapy [20]. Some other reported works in polymersomes as chemotherapeutic drug delivery shuttles include use of poly(ethylene oxide)-block-poly(ethyl ethylene) [21], poly(ethylene oxide)-block-poly(butadiene) [22], poly(ethylene oxide)-block-poly(caprolactone), and poly(lysine)-block-poly(phenyl alanine) [23].

Layer-by-layer assembly of polyspermine (a polymer based on polyamine) was studied in lung cancer efficacy by Hong et al. [24]. They prepared a copolymer of glycerol triacrylate (GT)-spermine (SPE), complexed with a tumor-suppressing DNA, and utilized the carrier in successful transfection of gene to the tumor cells which suppressed lung tumorigenesis through apoptosis with no organ toxicity. Tan et al. designed polyethylene glycol (PEG)-phosphatidylethanolamine (PE) nanomicelles and successfully administered orally quercetin drug (a flavonoid-based potential anticancer drug with poor water solubility and less cellular absorbability) to A549 lung cancer cell line in a xenografted model [25].

3.2 Dendrimer

Dendrimers are a family of nanosized, three-dimensional hyperbranched macromolecules, exhibiting a tree-like branching architecture (Fig. 3.2) [26, 27]. Beyond third generation, dendrimers assume spherical structures and occupy considerably smaller hydrodynamic volume than the conventional linear polymers of similar molecular weight [28]. Owing to the three-dimensional branching architecture,

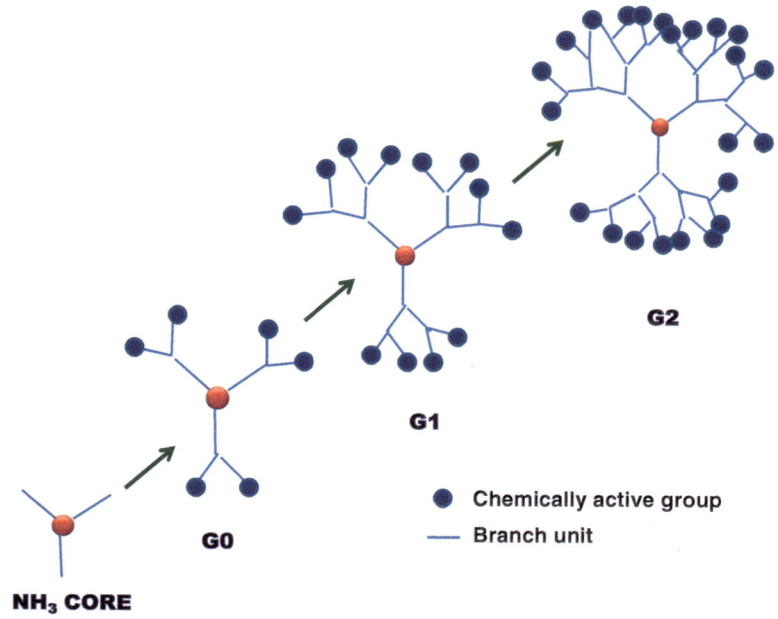

Fig. 3.2 Successive generations of a divergent dendrimer

low intrinsic viscosity, low polydispersity, high molecular weight, multivalency, and globular design with extensive feasibility for range of surface functionalization with bioactive molecules, they are well suited for the targeted delivery of antineoplastic drugs and imaging agents to the tumor cells. Drug molecules (DOX—an anthracycline antibiotics and lipophilic etoposide—a derivative of podophyllotoxin, used to treat any sort of cancer) may be loaded either into the interior of the dendrimers where lot of void spaces are present (via non-covalent encapsulation) or may be covalently attached to the surface-active functional groups [4]. Attached or encapsulated macromolecules are then delivered to the desired damaged cells by passive targeting (EPR effect) [29, 30]. Dendrimers are often used as non-viral vectors for effective gene transfection into the damaged cell nucleus by endocytosis. A pharmaceutical company named Starpharma Holdings Ltd. used a PAMAM dendrimer–DOX formulation to precisely deliver the PTX to lung-resistant tumors metastasized from breast cancers via intra-tracheal route and found that the formulation dramatically reduced the lung metastasis (Fig. 3.3) [31].

Dendrimers of poly(amidoamine) (PAMAM) [32], poly(glycerol succinic acid) (PGSLSA) [33], and poly(propylene imine) (PPI) [34] are extensively used to encapsulate antineoplastic drugs, interfering genes, and other macromolecules and to deliver them successfully to NSCLC carcinoma cells. Apart from the use of dendrimer NPs in drug/gene delivery, Liu et al. showed that NSCLC-targeting peptide ligands (sequence RCPLSHSLICY) and FITC (a fluorescent label) molecules-conjugated acetylated PAMAM dendrimer could be effectively used for both in vitro and in vivo targeting to

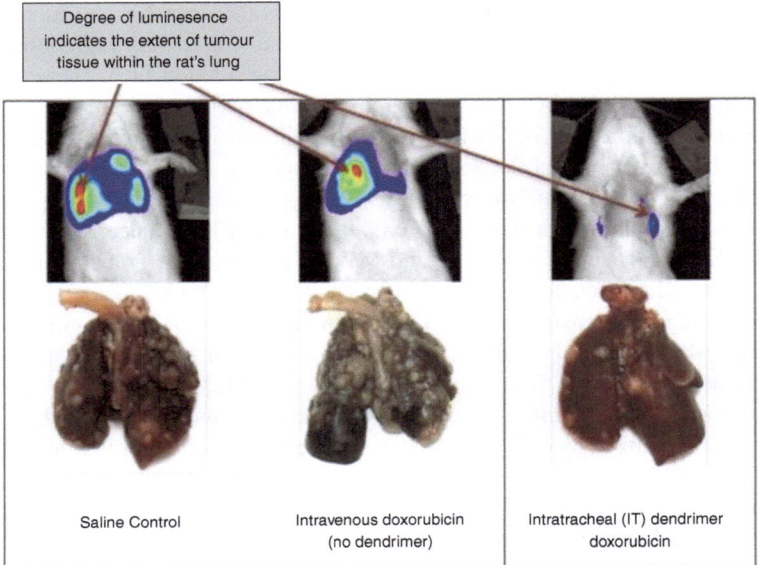

Degree of luminesence indicates the extent of tumour tissue within the rat's lung

Saline Control | Intravenous doxorubicin (no dendrimer) | Intratracheal (IT) dendrimer doxorubicin

Fig. 3.3 Showing results from the study in which rats with lung-resident tumors derived from breast cancer cells were treated with either saline (*left panel*), intravenous doxorubicin (*center panel*), or intratracheal dendrimer–doxorubicin (*right panel*). Both the gross appearance of the lungs and the bioluminescent images in the final stage revealed that the extent of lung metastasis was greatly reduced in the presence of dendrimer–doxorubicin complex as compared to the other treatments (*Credit* image courtesy by Starpharma Holdings Ltd., http://www.starpharma.com/news/150, Access date 21 Feb 2014)

and cellular imaging of NCI-H460 lung carcinoma cells without any unwanted accumulation of strong therapeutic agents in other parts of the body [35]. Recently, conjugation of dendrimers with gold NPs has attracted the area of cancer diagnosis and imaging. This is discussed later in details under the broad head of gold nanoparticles in Sect. 4.1.

3.3 Carbon Nanoparticles

Carbon NPs (CNPs) used in the realm of biomedicines is widely categorized as carbon nanotubes (CNTs) and carbon nanofibers (CNFs). CNTs are mostly used in lung cancer therapy. CNTs are hypothetically either made up of single layer of graphene sheet (single-walled carbon nanotube: SWNT with diameter of 0.4–2 nm) or concentric multiple layers of SWNTs (multiwalled carbon nanotube: MWNT with inner diameter of 1–3 nm and outer diameter of 2–100 nm), commonly held together through sp^2 bonds [36, 37]. The layers are rolled into a seamless cylinder that can be open ended or capped at the extremities with a buckyball, having a high aspect ratio with diameters as small as 1 nm and a length of several micrometers [38, 39]. CNTs are used as potential drug/gene delivery systems as they can bind to multitude of therapeutic and biologically active macromolecules owing to very high surface area and presence of large empty internal space for encapsulation [40]. However, pristine CNTs are insoluble in aqueous medium like human bloodstream. Hence, facile surface modifications of CNTs with hydrophilic and organic moieties enhance their dispersibility in aqueous phase and engineer the surface for adsorption or conjugation to targeting therapeutic macromolecules. A group headed by Adeli prepared a CNT-based formulation for the site-specific treatment of lung cancer. They oxidized MWCNTs and covalently functionalized MWCNTs with hyperbranched poly(citric acid) (PCA). They then conjugated MWCNTs–PCA with PTX molecules. The final complex being acid sensitive successfully released PTX at the tumor microenvironment of A549 cells through enzymatic hydrolysis in an in vitro study and exhibited reduced side effects than free taxols on healthy cells [41]. Often, CNTs are conjugated to various biopolymers like peptides, proteins, or nucleic acids. Thus, CNTs are easily internalized by cells by endocytosis due to EPR effect of tumor cells. Also CNTs exhibit high optical absorbance in the near-infrared (NIR) region and radiofrequency range and are thus effectively used in thermal ablation of cancerous cells [42–44]. Photoluminescence properties of CNTs make them attractive candidates in the field of cellular imaging and staging of cancer [42]. Thus, unique electrical, thermal, and spectroscopic properties of CNTs in a biological context offer further advances in the simultaneous detection, monitoring, and therapy of cancer diseases. Other variations of CNPs that are extensively used in biomedical field include carbon nanohorns (CNHs) and nanodiamonds (NDs) [40]. CNHs are constituted of SWNTs aggregated in a nanoscale globular arrangement similar to sea urchins or dahlias, capped with a cone-shaped cap of five carbon pentagon rings together with many carbon hexagons [45]. NDs are three-dimensional structures in which carbon atoms have sp^3 hybridization, as in diamonds, but the dimensions remain in the nanometer range [46]. Liu et al. used tricosane ($C_{23}H_{48}$) and pentadecane ($C_{15}H_{32}$)/dioctylphthalate ($C_{24}H_{38}O_4$)-modified SWNT-based biosensor to detect both polar (1,2,4-trimethylbenzene) and nonpolar (decane) lung cancer biomarking volatile organic compounds (VOCs) in human breathe with high sensitivity and selectivity [47]. Murakami and his group prepared high-density stabilized lipoprotein (HDL)–SWNTs complexes, treated human lung cancer cells (NCI-H460) with HDL–SWNTs, irradiated HDL–SWNT-treated cancer cells with 808-nm laser for 10 min and executed successful photodynamic killing of cancerous cells by generation of highly reactive ($^1O_2$) species [48]. Podesta [49] and his team efficiently used amino-functionalized multiwalled carbon nanotubes (MWNT-NH) to deliver siRNA sequences to human lung carcinoma cells and induced mitotic apoptosis which eventually reduced tumor growth in xenografted animals. Ajima et al. [50] reported successful delivery of cisplatin drugs from CNHs/SWNHs to NCI-H460 cell lines without any cytotoxic effects. In another study, Liu et al. treated A549 human carcinoma cells with ND-loaded PTX drug and noticed both mitotic arrest and apoptosis of A549 cells due to selectively delivery of the antineoplastic drugs to the abnormal cells. Despite the high tunable properties of various forms of CNPs which make them attractive in biomedical applications, CNPs are not used in any clinical trials.

There is strong evidence that CNPs cause oxidative damage to cellular membranes [51, 52]. Thus, further research and preclinical trials are required before CNP-based biomedicine could be commercialized for in vivo studies in human.

References

1. Rao JP, Geckeler KE (2011) Polymer nanoparticles: preparation techniques and size-control parameters. Prog Polym Sci 36:887–913
2. Jung J, Park SJ, Chung HK, Kang HW, Lee SW, Seo MH, Park HJ, Song SY, Jeong SY, Choi EK (2012) Polymeric nanoparticles containing taxanes enhance chemoradiotherapeutic efficacy in non-small cell lung cancer. Int J Radiat Oncol Biol Phys 84:e77–e83
3. Sengupta S, Eavarone D, Capila I, Zhao G, Watson N, Kiziltepe T, Sasisekharan R (2005) Temporal targeting of tumour cells and neovasculature with a nanoscale delivery system. Nature 436:568–572
4. Benfer M, Kissel T (2012) Cellular uptake mechanism and knockdown activity of siRNA-loaded biodegradable DEAPA-PVA-g-PLGA nanoparticles. Eur J Pharm Biopharm 80:247–256
5. Fischer D, Li Y, Ahlemeyer B, Krieglstein J, Kissel T (2003) In vitro cytotoxicity testing of polycations: influence of polymer structure on cell viability and hemolysis. Biomaterials 24:1121–1131
6. Zhou X, Li X, Gou M, Qiu J, Li J, Yu C, Zhang Y, Zhang N, Teng X, Chen Z, Luo C, Wang Z, Liu X, Shen G, Yang L, Qian Z, Wei Y, Li J (2011) Antitumoral efficacy by systemic delivery of heparin conjugated polyethylenimine–plasmid interleukin-15 complexes in murine models of lung metastasis. Cancer Sci 102:1403–1409
7. Jere D, Jiang HL, Kim YK, Arote R, Choi YJ, Yun CH, Cho MH, Cho CS (2009) Chitosan-graft-polyethylenimine for Akt1 siRNA delivery to lung cancer cells. Int J Pharm 378:194–200
8. Rajagopal K, Christian DA, Harada T, Tian A, Discher DE ()2010 Polymersomes and worm-like micelles made fluorescent by direct modifications of block copolymer amphiphiles. Int J Polym Sci 10
9. Kim Y, Dalhaimer P, Christian DA, Discher DE (2005) Polymeric worm micelles as nano-carriers for drug delivery. Nanotechnology 16:S1–S8
10. Cai S, Vijayan K, Cheng D, Lima E, Discher D (2007) Micelles of different morphologies-advantages of worm-like filomicelles of PEO-PCL in paclitaxel delivery. Pharm Res 24:2099–2109
11. Van Domeselaar GH, Kwon GS, Andrew LC, Wishart DS (2003) Application of solid phase peptide synthesis to engineering PEO-peptide block copolymers for drug delivery. Colloids Surf B 30:323–334
12. Lee J, Cho EC, Cho K (2004) Incorporation and release behavior of hydrophobic drug in functionalized poly(D, L-lactide)-block-poly(ethylene oxide) micelles. J Controlled Release 94:323–335
13. Geng Y, Discher DE (2006) Visualization of degradable worm micelle breakdown in relation to drug release. Polymer 47:2519–2525
14. Geng Y, Dalhaimer P, Cai S, Tsai R, Tewari M, Minko T, Discher DE (2007) Shape effects of filaments versus spherical particles in flow and drug delivery. Nat Nanotechnol 2:249–255
15. Discher BM, Bermudez H, Hammer DA, Discher DE, Won YY, Bates FS (2002) Cross-linked polymersome membranes: vesicles with broadly adjustable properties. J Phys Chem B 106:2848–2854
16. Ghoroghchian PP, Frail PR, Susumu K, Blessington D, Brannan AK, Bates FS, Chance B, Hammer DA, Therien MJ (2005) Near-infrared-emissive polymersomes: self-assembled soft matter for in vivo optical imaging. Proc Natl Acad Sci USA 102:2922–2927

17. Discher DE, Ahmed F (2006) Polymersomes. Annu Rev Biomed Eng 8:323–341
18. Ghoroghchian PP, Li G, Levine DH, Davis KP, Bates FS, Hammer DA, Therien MJ (2006) Bioresorbable vesicles formed through spontaneous self-assembly of amphiphilic poly(ethylene oxide)-block-polycaprolactone. Macromolecules 39:1673–1675
19. Waterhouse DN, Tardi PG, Mayer LD, Bally MB (2001) A comparison of liposomal formulations of doxorubicin with drug administered in free form: changing toxicity profiles. Drug Saf 24:903–920
20. Ahmed F, Pakunlu RI, Srinivas G, Brannan A, Bates F, Klein ML, Minko T, Discher DE (2006) Shrinkage of a rapidly growing tumor by drug-loaded polymersomes: pH-triggered release through copolymer degradation. Mol. Pharmaceutics 3:340–350
21. Discher BM, Won Y-Y, Ege DS, Lee JCM, Bates FS, Discher DE, Hammer DA (1999) Polymersomes: tough vesicles made from diblock copolymers. Science 284:1143–1146
22. Bermudez H, Brannan AK, Hammer DA, Bates FS, Discher DE (2002) Molecular weight dependence of polymersome membrane structure, elasticity, and stability. Macromolecules 35:8203–8208
23. Sun J, Chen X, Deng C, Yu H, Xie Z, Jing X (2007) Direct formation of giant vesicles from synthetic polypeptides. Langmuir 23:8308–8315
24. Hong SH, Kim JE, Kim YK, Minai-Tehrani A, Shin JY, Kang B, Kim HJ, Cho CS, Chae C, Jiang HL, Cho MH (2012) Suppression of lung cancer progression by biocompatible glycerol triacrylate-spermine-mediated delivery of shAkt1. Int J Nanomed 7:2293–2306
25. Tan BJ, Liu Y, Chang KL, Lim BK, Chiu GN (2012) Perorally active nanomicellar formulation of quercetin in the treatment of lung cancer. Int J Nanomed 7:651–661
26. Medina SH, El-Sayed MEH (2009) Dendrimers as carriers for delivery of chemotherapeutic agents. Chem Rev 109:3141–3157
27. Hari BNV, Kalaimagal K, Porkodi R, Gajula PK, Ajay JY (2012) Dendrimer: globular nanostructured materials for drug delivery. Int J PharmTech Res 4:432–451
28. Sampathkumar S-G, Yarema KJ (2007) Dendrimers in cancer treatment and diagnosis, nanotechnologies for the life sciences. Wiley-VCH Verlag GmbH & Co. KGaA, Weinheim
29. Goya GF, Grazu V, Ibarra MR (2008) Magnetic nanoparticles for cancer therapy. Curr Nanosci 4:1–16
30. Vicent MJ, Duncan R (2006) Polymer conjugates: nanosized medicines for treating cancer. Trends Biotechnol 24:39–47
31. http://www.starpharma.com/news/150
32. Rahbek UL, Nielsen AF, Dong M, You Y, Chauchereau A, Oupicky D, Besenbacher F, Kjems J, Howard KA (2010) Bioresponsive hyperbranched polymers for siRNA and miRNA delivery. J Drug Target 18:812–820
33. Morgan MT, Nakanishi Y, Kroll DJ, Griset AP, Carnahan MA, Wathier M, Oberlies NH, Manikumar G, Wani MC, Grinstaff MW (2006) Dendrimer-encapsulated camptothecins: increased solubility, cellular uptake, and cellular retention affords enhanced anticancer activity in vitro. Cancer Res 66:11913–11921
34. Taratula O, Garbuzenko OB, Kirkpatrick P, Pandya I, Savla R, Pozharov VP, He H, Minko T (2009) Surface-engineered targeted PPI dendrimer for efficient intracellular and intratumoral siRNA delivery. J Controlled Release 140:284–293
35. Liu J, Chu L, Wang Y, Duan Y, Feng L, Yang C, Wang L, Kong D (2011) Novel peptide-dendrimer conjugates as drug carriers for targeting nonsmall cell lung cancer. Int J Nanomed 6:59–69
36. Bianco A, Kostarelos K, Prato M (2005) Applications of carbon nanotubes in drug delivery. Curr Opin Chem Biol 9:674–679
37. Heister E, Neves V, Tilmaciu C, Lipert K, Beltran VS, Coley HM, Silva SRP, McFadden J (2009) Triple functionalisation of single-walled carbon nanotubes with doxorubicin, a monoclonal antibody, and a fluorescent marker for targeted cancer therapy. Carbon 47:2152–2160
38. Surendiran A, Sandhiya S, Pradhan SC, Adithan C (2009) Novel applications of nanotechnology in medicine. Indian J Med Res 130:689–701
39. Meng L, Zhang X, Lu Q, Fei Z, Dyson PJ (2012) Single walled carbon nanotubes as drug delivery vehicles: targeting doxorubicin to tumors. Biomaterials 33:1689–1698

40. Bianco A, Kostarelos K, Prato M (2008) Opportunities and challenges of carbon-based nano-materials for cancer therapy. Expert Opin Drug Deliv 5:331–342

41. Sobhani Z, Dinarvand R, Atyabi F, Ghahremani M, Adeli M (2011) Increased paclitaxel cytotoxicity against cancer cell lines using a novel functionalized carbon nanotube. Int J Nanomed 6:705–719

42. Bachilo SM, Strano MS, Kittrell C, Hauge RH, Smalley RE, Weisman RB (2002) Structure-assigned optical spectra of single-walled carbon nanotubes. Science 298:2361–2366

43. Madani SY, Tan A, Dwek M, Seifalian AM (2012) Functionalization of single-walled carbon nanotubes and their binding to cancer cells. Int J Nanomed 7:905–914

44. Kam NWS, O'Connell M, Wisdom JA, Dai H (2005) Carbon nanotubes as multifunctional biological transporters and near-infrared agents for selective cancer cell destruction. Proc Natl Acad Sci USA 102:11600–11605

45. Iijima S, Yudasaka M, Yamada R, Bandow S, Suenaga K, Kokai F, Takahashi K (1999) Nano-aggregates of single-walled graphitic carbon nano-horns. Chem Phys Lett 309:165–170

46. Ferro S (2002) Synthesis of diamond. J Mater Chem 12:2843–2855

47. Liu FL, Xiao P, Fang HL, Dai HF, Qiao L, Zhang YH (2011) Single-walled carbon nanotube-based biosensors for the detection of volatile organic compounds of lung cancer. Physica E 44:367–372

48. Murakami T, Nakatsuji H, Inada M, Matoba Y, Umeyama T, Tsujimoto M, Isoda S, Hashida M, Imahori H (2012) Photodynamic and photothermal effects of semiconducting and metal-lic-enriched single-walled carbon nanotubes. J Am Chem Soc 134:17862–17865

49. Podesta JE, Al-Jamal KT, Herrero MA, Tian B, Ali-Boucetta H, Hegde V, Bianco A, Prato M, Kostarelos K (2009) Antitumor activity and prolonged survival by carbon-nanotube-mediated therapeutic siRNA silencing in a human lung xenograft model. Small 5:1176–1185

50. Ajima K, Yudasaka M, Murakami T, Maigne A, Shiba K, Iijima S (2005) Carbon nanohorns as anticancer drug carriers. Mol Pharm 2:475–480

51. Cancino J, Paino IMM, Micocci KC, Selistre-de-Araujo HS, Zucolotto V (2013) In vitro nanotoxicity of single-walled carbon nanotube-dendrimer nanocomplexes against murine myoblast cells. Toxicol Lett 219:18–25

52. Oberdorster E (2004) Manufactured nanomaterials (fullerenes, C60) induce oxidative stress in the brain of juvenile largemouth bass. Environ Health Perspect 112:1058–1062

Chapter 4
Synthetic (Inorganic) Nanoparticles Based Lung Cancer Diagnosis and Therapy

Abstract Synthetically designed inorganic nanoparticles exhibit few striking intrinsic properties which enhance their therapeutic value over the natural and synthetic organic nanoparticles. In the previous chapters, various nanoparticles as discussed were mainly used as delivery shuttles for the delivery of macromolecules at the target locations. Undoubtedly, natural and organic nanoparticles are highly biocompatible, yet they fail to circumvent the growing mortality rate of lung cancer. Hence, inorganic nanoparticles if properly designed may improve lung cancer therapy in many ways. This chapter illustrates various magnificent properties of individual inorganic nanoparticles and explains broadly how those properties may be utilized for successful lung cancer therapy apart from just being used as delivery systems.

Keywords Gold nanoparticle · Magnetic nanoparticle · Quantum dot · Silica nanoparticle · Lanthanide nanoparticle · Early detection of lung cancer · Cellular imaging · Active targeted therapy · Hyperthermia

4.1 Gold Nanoparticles

Gold NPs (GNPs) are the most commonly employed inorganic NPs for cancer diagnostics and drug delivery. Gold is a noble metal. Colloidal gold has some unique properties which make them ideal for targeting therapy. Noble metal NPs exhibit high surface-to-volume ratio, broad optical properties, biological inertness, resistance to corrosion, low toxicity (as when the particle degrades, units are nontoxic and either easily cleared from the body by the renal system or incorporated into metabolic pathways), and good antimicrobial efficacy (even in very small concentration, against bacteria, viruses, and other eukaryotic microorganisms, a recent research found that GNPs can destroy DNA or cell wall of germs and bacteria by shrinkage of the cytoplasm membrane or its detachment from the cell wall). Alternatively, gold ions might interact with the S–H bonds of the biological

© The Author(s) 2015
A. Bandyopadhyay et al., *Nanoparticles in Lung Cancer Therapy - Recent Trends*,
SpringerBriefs in Molecular Science, DOI 10.1007/978-81-322-2175-3_4

proteins of microorganisms and inactivate them. Upon binding of GNPs to DNA, DNA molecules become condensed and lose their ability to replicate, which may be the main mechanism by which GNPs inhibit the bacterial replications [1]) and no cytotoxicity on epidermal cells. Moreover, GNPs are easy to synthesize and their tunable surface functionalization hold pledge in the clinical field. Even the GNPs present highly size-tunable optical properties like strong surface plasmon resonance (SPR) effect, i.e., produce bright colors owing to high absorption and scattering cross section in GNPs. The SPR effect of GNPs may be easily tuned to the wavelengths where is least blood and tissue attenuation, i.e., in the "biological window" (650–900 nm) according to their shape (e.g., nanoparticles, nanoshells, nanorods), size (e.g., 1–100 nm), composition (e.g., core/shell or alloy noble metals), and dielectric strength of the surrounding medium [2, 3]. SPR in GNPs mainly occurs due to the interaction between surface conduction electrons and applied electromagnetic waves, thereby producing an amplified coherent resonance which is basically tuned according to our need in biological systems [4]. Thus the SPR effect of GNPs which depends on size and shape of GNPs is visually represented in Fig. 4.1 [5].

The SPR phenomenon in GNPs generally exists at the interface of two media: GNPs (a conductor) and biological environment (a dielectric medium). This phenomenon obviously makes GNPs suitable as powerful imaging labels, contrast agents (CAs), and sensors and is effectively used in photothermal applications under native tissues [6, 7]. Also, GNPs are highly reactive, may readily bind with a variety of negatively charged molecules including inorganic labeled antibodies, anions, peptides, RNA, and DNA, and thereby improve in vivo imaging. In fact, gold has a strong affinity to thiols, disulfides, phosphines, and amine functionalities (according to HSAB principle, gold being a soft acid is known to bind strongly with soft aforementioned bases [8]). Such affinity enables surface modifications to GNPs relatively easy through Au–S and Au–N bonding with targeting agents and/or chemotherapeutics that possess these functionalities [9, 10]. GNPs

Fig. 4.1 Size-dependent SPR effect of GNPs (*Credit* Image courtesy by http://en.wikipedia.org/wiki/Colloidal_gold, Access date 15 Mar 2013)

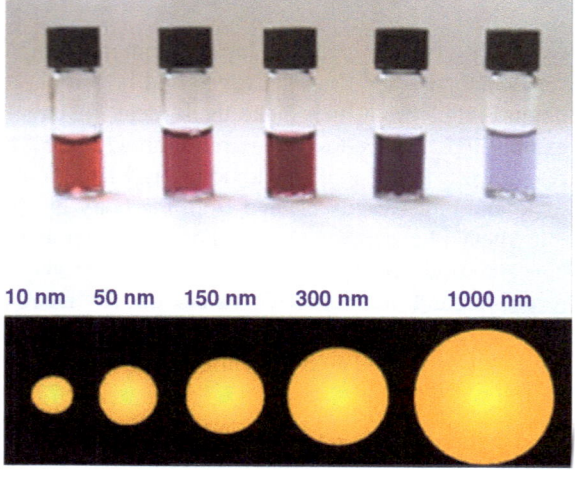

are also compatible with biopolymers [e.g., polyethylene glycol (PEG)] and form complexes with the latter to prolong their in vivo circulation time for drug and gene delivery applications. GNPs can efficiently convert light or radio frequencies into heat, thus enabling the thermal ablation of targeted cells. If GNPs are properly conjugated with desirable target moieties, then they may be easily engulfed by the cancer cells into their nucleus. When exposed to infrared laser (of wavelength harmless to normal human tissue), GNPs get heated up by the absorption of the light which in turn heat the host cancerous cells and kill them through localized heating. Last but not the least, gold (I and III) compounds themselves demonstrated anticancer properties. They can also enhance the antitumor activities of the encapsulated antitumor compounds [9, 11]. It is thus understood that why GNPs are considered to be the king in the realm of nanobiomedicine. GNPs are applied either naked or as bioconjugates for early detection, diagnosis, imaging, targeting, and treatment of malicious tumors. Some of the recently reported works using GNPs in the territory of bronchogenic carcinomas are highlighted as follows. One of the fastest methods of screening of lung cancer at an early stage is breath testing using electronic nose (Fig. 4.2).

Human breath contains approximately 200 VOCs on an average, mostly in picomolar concentration (i.e., 10–12 mol/L) [12]. Gas chromatography/mass spectrometry (GC-MS) studies have shown that various VOCs in a healthy breath normally appear at 1–20 ppb. However, it is raised to an anomalous concentration (10–100 ppb) in breath of lung cancer patients [13–15]. Some of the VOCs are recognized as potential biomarkers (indicators of a biological state of disease) for lung cancer. For instance, alkanes, methylated alkanes, and acetaldehydes are detected at an abnormally high concentration in lung cancer breath [12, 16]. Alkanes,

Fig. 4.2 Image of breath testing using electronic nose composed of GNPs (*Credit* Image courtesy by Emily Anthes, http://www.scientificamerican.com/article/electronic-noses-could-make-diseases-something-to-sniff-at/, Access date 02 Mar 2013)

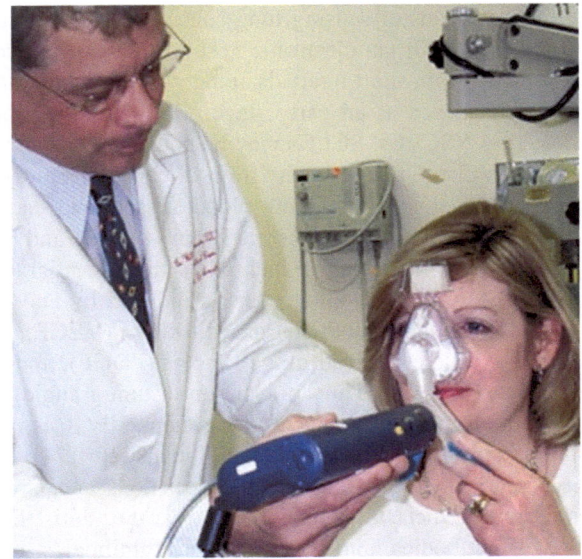

aldehydes, and ethane and pentane volatiles are generated during lipid peroxidation of unsaturated fatty acids by reactive oxygen species (ROS) [17, 18]. This difference in VOC content between breaths of normal human and patients suffering from lung cancer is utilized to detect lung cancer at an early stage. So far, GC-MS [13, 15], ion flow tube mass spectrometry [19], laser absorption spectrometry [20], infrared spectroscopy [21], polymer-encapsulated surface acoustic wave sensors [14], and coated quartz crystal microbalance sensors [22] are extensively used to determine the composition of VOCs in human breath. However, the aforementioned techniques suffer from shortcomings such as high cost, slow and tedious process, complex instrumentation, requirement of skilled analysts, and pretreatment of biomarkers to increase the concentration to a level so that it becomes detectable. Along with preconcentration, predehumidification is also done to improve the sensitivity of the tests for early detection. Even an array of 10 sensors based on organically functionalized CNTs failed to provide the satisfactory results without pretreatment. But it is not possible to obtain complete composition of the VOCs by these techniques. They skip certain VOCs which are present in minor quantities. Thus, this causes loss of lives. Recently, Peng et al. [16] designed an array of nine different chemiresistors based on functionalized 5-nm GNPs to detect different percentages of VOCs in the infected exogenous (those inhaled or absorbed through the skin and then inhaled) and endogenous breath (produced by different biochemical processes). GNP-based sensors are highly sensitive owing to their inherent SPR property. They functionalized GNPs with nine organic capping agents such as dodecanethiol, decanethiol, 1-butanethiol, 2-ethylhexanediol, hexanethiol, tert-dodecanethiol, 4-methoxy-toluenethiol, 2-mercaptobenzoxazole, and 11-mercapto-1-undecanol and used them to sense different cancer biomarking VOCs in the infected breath. The sensitivity of GNP-based sensors in sensing lung cancer biomarkers was hardly affected by the presence of moisture in breath. Thus, it is a very effective means of sensing lung cancer at an early stage at reasonable costs. They collected breath via electronic nose of GNPs even from very severe patients, also repeated within short intervals along with clinical treatments and obtained very useful results even at an early stage. GNP-based sensors are, however, suitable for detecting NSCLCs. SCLCs may be detected at a later stage. Angiogenesis is a very common phenomenon in any form of cancer. Angiogenesis, by definition, is a process of development of new blood vessels and capillaries from the pre-existing ones. These extra blood vessels supply oxygen and other essential nutrients to the fast-growing tumor cells which allow them to migrate and metastasize to different organs. Any tumor angiogenesis is induced by endothelial cell-specific mitogens, such as vascular endothelial growth factor (VEGF), basic fibroblast growth factor (bFGF), platelet-derived growth factor (PDGF), and transforming growth factor-β (TGF-β) [8, 9, 23]. Thus, prevention of tumor angiogenesis may prove to be a cornerstone to prevent cancer. Among them, VEGF is mostly responsible for sustaining growth in both normal and malignant cells [24]. Increased expression of VEGF is prominent in NSCLC-based cancer cells and is often associated with risks of recurrence, metastasis, and death. Various anti-VEGF angiogenic agents (anti-VEGF antibodies, anti-VEGF receptor antibodies, and VEGF receptor inhibitors) are available in the market, but unfortunately, many of them have proved to be toxic. Recently, GNPs have proved to be a potentially biocompatible antiangiogenic agents. GNPs may selectively bind to heparin domains, inhibit heparin-binding growth factors such as VEGF and bFGF, and thereby eliminate the interaction between VEGF and its receptor [8]. Hyperthermia is the process of application of heat to kill tumor cells at a temperature above 40 °C. Higher temperature causes necrosis (premature cell death) of the living cells, denatures enzymes, brings about structural and functional changes in DNA or RNA, releases cellular content due to rupture of cell membrane, and finally results in the death of the cells. However, hyperthermia lacks specificity, i.e., kills both tumor and healthy cells simultaneously. GNPs when used as the probe may selectively differentiate healthy and malignant cells while killing. GNPs are conjugated with desirable antitumor moieties and are thus selectively engulfed into the nucleus of malignant cells. Finally, GNPs owing to their strong SPR absorption may convert light or radio frequencies to heat which ultimately damages the fast-growing tumor cells. However, radio frequencies (13.56 MHz) are more promising for native tissue penetration [25]. Cheng et al. [26] in a preclinical trial used HER2 antibody-modified silica@Au nanoshells, hollow Au/Ag nanospheres, and Au nanorods to kill malignant A549 cells through photothermal effect of NIR light. In their study, silica@Au nanoshells exhibited the best photothermal ablation of malignant tissues at a minimum dosage. When GNPs were locally overheated with short laser pulses, they become hot so quickly that explosion took place. This explosion mainly occurred due to rapid evaporation of the very thin volume of the surrounding medium which produced a gradually expanding vapor nanobubble. On reaching critical size, the nanobubble collapsed with an explosion within nanoseconds. This obviously led to the death of malignant cells as the entire arrangement as they precisely located silica@Au nanoshells within the affected cells by accurate design of the GNPs with targeting moieties and protecting sheath [25]. Immunoassay defines biochemical assay of specific immune reaction between an antigen and an antibody. Conventional immunoassay includes radio immunoassay (antigen labeled with radioactive elements), enzyme-linked immunosorbent assay (ELISA) (enzyme as label), or other variations including detectable dyes, fluorescent materials, chemiluminescent dyes, metal chelates, and magnetic particles etc when used as antigen labels. However, conventional immunoassay suffers from major drawbacks. The processes are complicated, multistage, tedious, expensive, and difficult to employ in situ. Electrochemical (amperometric) immunoassay is gaining impetus to ensure quick assay. However, antigens and antibodies are electrochemically inert. So often, they are labeled with probes of GNPs, CNTs, or quantum dots (QDs) [27]. A typical immunoassay process involves the detection of produced hydrogen peroxide (H_2O_2) owing to the oxidation of glucose, catalyzed by glucose oxidase (GOx), in an enzymatic reaction [27, 28]. Now this produced H_2O_2 may be sensed by its reduction or oxidation at a solid electrode. However, anodic determination of H_2O_2 is unpractical because it requires a very high oxidation potential and is also interfered by other oxidizable substances (ascorbic acid, uric acid, etc.) present in the samples. On the contrary, cathodic determination of H_2O_2 using an electrocatalyst like horseradish peroxidase (HRP) is feasible due to low oxidation potential. At low oxidation potential, interfering species are electrochemically inactive and thus provide a suitable environment for biosensing. Although HRP is a potential catalyst in the reduction of H_2O_2, it is found to adversely affect the stability of biosensors owing to denaturation [28]. HRPs are also expensive. In an attempt to substitute HRP, hydrazine is explored in biosensors. Hydrazine is a strong reducing agent and thus may be used as an electrocatalyst in immunoassay. Now that Annexin II and MUC5AC are potential biomarkers (antigens) in patients with lung cancer. Both are overexpressed in neoplastic cells including A549 cells. Therefore, early detection of these biomarkers may save life. This demands the design of a stable and sensitive amperometric immunosensor. Development of GNP–dendrimer conjugate served the purpose [27]. Kim et al. in the primary step of their novel design for a preclinical trial electrodeposited GNPs onto the glassy carbon electrode (GCE) and simultaneously polymerized a terthiophene monomer having a carboxylic acid group (5,2′:5′,2″-terthiophene-3′-carboxylic acid) which stabilized the nanoparticles. In the poly-TTCA–GNP electrode, they activated the carboxylic acid groups of poly-TTCA layer by EDC and finally used to immobilize an amine-terminated dendrimer via amide bond formation. Again, they electrodeposited GNPs on the dendrimer/poly-TTCA/GNP-modified electrode. In the following step, they activated the amine groups of dendrimer by glutaraldehyde and linked to hydrazine sulfate. Finally, they treated the entire system with anti-Annexin II antibodies and completed the formation of immunosensing probe. A schematic representation of the entire process design of the immunosensing probe is shown in Fig. 4.3. In the collected samples, they labeled the Annexin II antigen with the enzyme, GOx, and then introduced into a glucose solution. When the anti-Annexin II antibodies treated, hydrazine sulfate-based dendrimer/poly-TTCA/GNP-modified electrode came in contact with GOx-labeled Annexin II–glucose samples, and Annexin II was easily detected via an enzymatic reaction between Annexin II antigen and anti-Annexin II antibodies. The same procedure was followed to sense MUC5AC by the developed immunosensor via an enzymatic antigen–antibody reaction.

Dendrimers, often due to their multitasking ability, have proved to be an effective tool in the realm of cancer diagnosis as mentioned in Sect. 3.2. An interesting study shows that biocompatible hydroxyl- or acetyl-functionalized

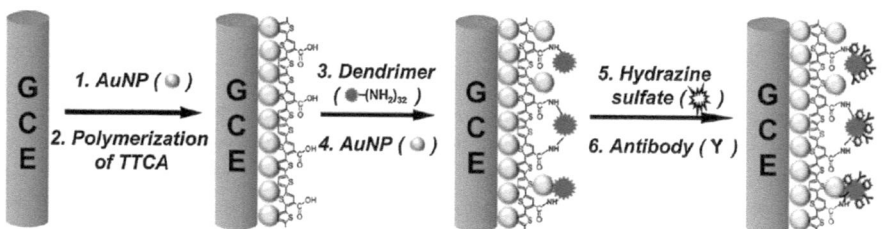

Fig. 4.3 Schematic representation of immunosensor fabrication "Reprinted (adapted) with permission from Kim et al. [27]. Copyright (2009) Elsevier"

dendrimer-entrapped GNPs (Au DENPs) may play a vital role in cancer cell-targeting and simultaneous imaging [29]. Here, the general mechanism of inhibition of cancer cell proliferation is based on Trojan Horse trickery (trickery of surprise killer). The principle underlines that all living cells require folic acid (FA) to replicate. However, cancer cells have extraordinary appetite for the same, and thereby, any carcinoma cell lines exhibit more docking sites of FA receptor. Thus, when Au DENPs linked with FA and fluorescein isothiocyanate (FI) molecules are introduced into a cancer patient, they may specifically bind to KB cells of the human epithelial carcinoma cell line due to the presence of overexpressed FA receptors and finally get internalized through receptor-mediated endocytosis (Fig. 4.4a). FI molecules owing to their fluorescence property enable cellular imaging when excited externally with a light of particular wavelength (Fig. 4.4b). In a similar study, Arvizo et al. [30] linked the Au DENPs to anti-EGFR antibodies and specifically targeted to overexpressed EGFR (epidermal growth factor receptors) in NSCLC-type lung cancer cells and enabled early lung cancer diagnosis .

GNPs are also used to deliver various therapeutic molecules at the target locations. Water-soluble chemotherapeutic drugs such as DOX and methotrexate (MTX) generally have poor tumor retention ability which significantly reduces drug efficacy. Chen et al. [31] showed that when MTX was conjugated to GNPs and targeted to LLC cells of a xenografted mouse accumulated in high concentration at the target locations and thus proved to be very effective in the treatment of cancer.

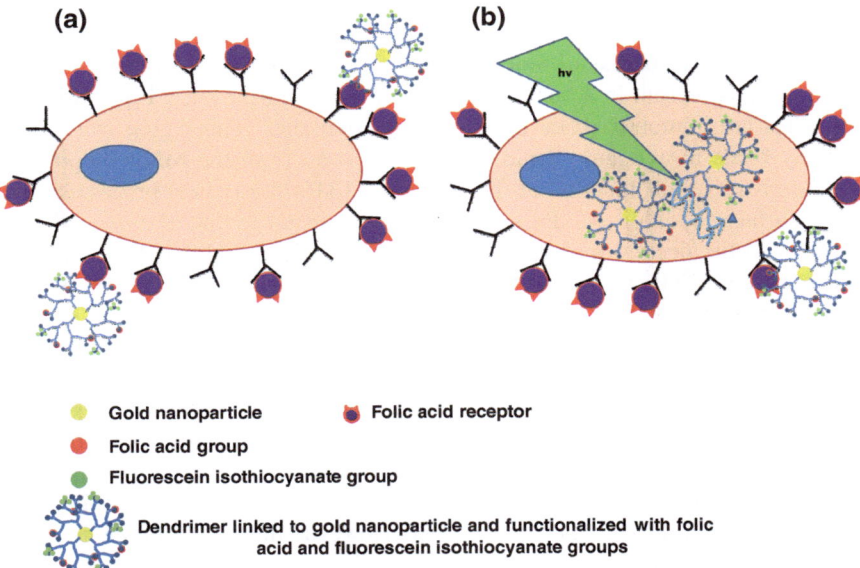

Fig. 4.4   a Location of FA- and FI-functionalized Au DENPs in the surrounding of a tumor cell by Trojan Horse trickery. b Cellular imaging by the excitation of FA/FI-functionalized Au DENPs accumulated in the tumor cell

4.2 Magnetic Nanoparticles

Magnetic nanoparticles (MNPs) are another important class of inorganic NPs chosen to widen the arena of cancer therapies, especially pulmonary cancer. MNPs have a wide range of controllable sizes from a few nanometers to micrometers. Their small sizes are comparable to cells, viruses, proteins, genes, or other biological entities which make them suitable for circulation in bloodstream and cellular uptake. Tagging of MNP surface with biological units converts them to potential vectors for delivering essential tumor-suppressing genes or drugs to the specific cells where they are supposed to be deficient. Apart from the advantages offered by any other NPs, MNPs have additional properties which make them ideal candidate for active (guided by external field) as well as passive targeting (guided by EPR effect) to cancerous cells. Due to the inherent magnetic properties of MNPs, their action within the patients' body may be controlled externally by magnetic field. They easily penetrate the deep underlying damaged tissues without any concomitant loss of therapeutic efficiency of drug/genes loaded onto MNPs. So far, biomedical application of MNPs relies on three vital principles [32]:

1. Active targeting of MNPs to desirable target locations (cancer cells) under controlled external magnetic field gradients
2. Magnetic hyperthermia owing to hysteresis loss
3. Development of contrast media owing to the inherent magnetic moments of MNPs

Today, magnetic separation of biological entities by labeling cell with biocompatible MNPs is a successful tool in eliminating the difficulties of analysis of the damaged cells present in a complicated native environment [33]. Typically, when antibodies are immobilized on MNPs, recognition sites (Fab region from where antibody binds to antigen) should be oriented away from the MNP support surfaces to preserve full function of the antibodies [34]. In a typical process, at first, stable surface-modified MNPs such as iron oxide magnetite (Fe_3O_4) NPs are tagged with the desired biological entity. These surface-modified MNPs are just biocompatible MNPs attached to specific antibodies or other biomacromolecules such as hormones or FA (suitable for targeting lung cancer receptor cells rich in FA receptors) and are generally shielded by coating of dextran, PVA, phospholipids, or other polymers. By lock-and-key mechanism, they targeted such MNP labels to specific antigens and thereby tagged defected cells. Then, they separated the magnetically labeled biological entities from the native solution by pumping a fluid mixture through a region with magnetic field gradient (say a column loosely packed with a magnetizable matrix of steel wool or beads). This immobilized the magnetic material through a magnetic force. However, there were problems of settling or adsorption of magnetically tagged entities onto the surface of the matrix. Use of oriented field gradient system such as quadrupole arrangement creates a magnetic gradient radially outward from the center of the flow column which prevented settling or adsorption of the magnetic analytes as shown in Fig. 4.5. This

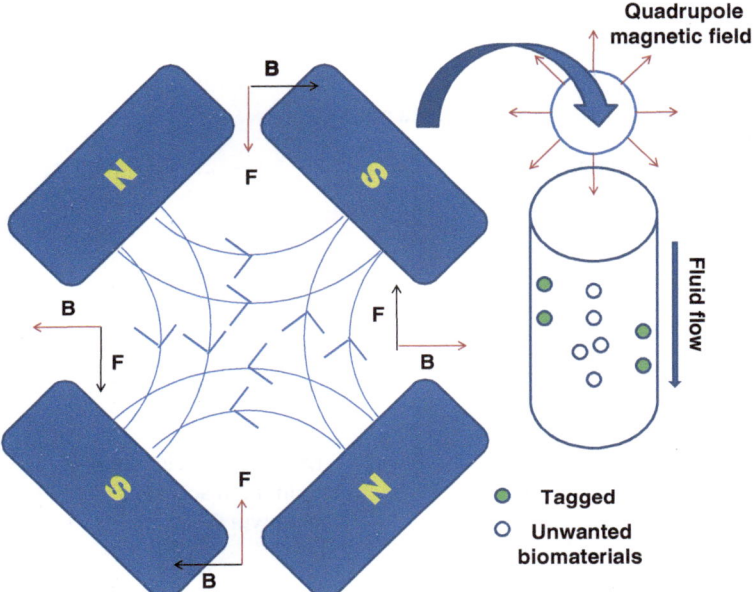

Fig. 4.5 A method of magnetic separation, in which an annular column containing a flowing solution of magnetically tagged (*filled circle*) and unwanted (*open circle*) biomaterials is placed within a set of magnets arranged in quadrature. Image is showing transverse cross section of the four magnets with the resulting magnetic field lines. Under the action of quadrature magnetic field gradient, the tagged particles move to the column walls, where they are held until the field is removed and they are recovered by flushing through with water. The central core of the column is made of non-magnetic material to avoid complications due to the near-zero field gradients

type of application of nanobiotechnology has improved immunoassay of pure analytes of pulmonary cancer at an early stage. Often, the magnetic separation technique is combined with optical sensing in ELISA. It just utilizes fluorescent enzymes to facilitate imaging for a clear development of idea of the disease.

Wang et al. [35] described successful detection of micrometastasis in lung cancer using MNPs. They conjugated MNPs with epithelial tumor cell markers pancytokeratin and very efficiently isolated circulating tumor cells (CTCs) from patients' blood by applying a magnetic field. MNPs have also attracted the field of artificial magnetic hyperthermia process to kill malignant cells selectively from a cluster of healthy cells (Fig. 4.6) [5]. Magnetic hyperthermia is a noninvasive therapeutic approach for lung cancer. Functionalized biocompatible MNPs are at first located inside the tumor cells and then heated by applying alternating magnetic field (AMF) of sufficient field strength or frequency which in turn kills the malignant cells when temperature crosses the therapeutic threshold of 42 °C, at least for 30 min [5].

Heat is generally developed in MNPs due to magnetic hysteresis loss (Neel or Brownian relaxation) under AMF. However, the field strength or frequency needed to heat an MNP is found to be beyond the tolerable limit of mammalian cells. Generally, high field strength is required for heating these ligand-targeting MNPs

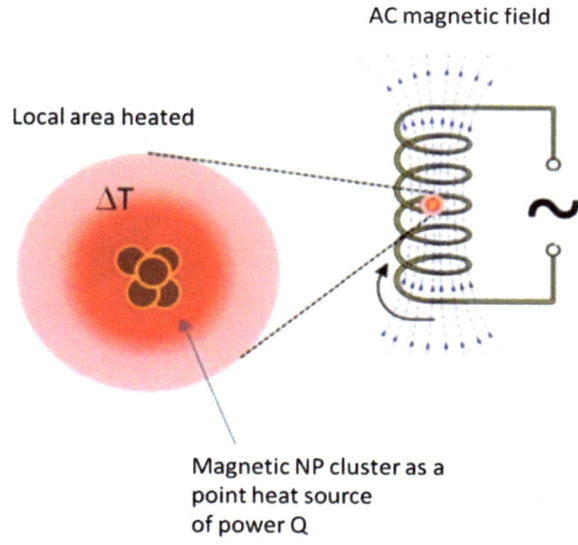

Fig. 4.6 Magnetic hyperthermia of MNPs causing local heating in tumors under alternating magnetic field (*Credit* Image courtesy by http://inl.int/projects/43, Access date: 15 May 2014)

as the cells need to be heated against the natural cooling mechanism by blood flow and tissue perfusion. So far, it is not an accepted clinical trial. However, with the improvement in ligand-specific localization and distribution of MNPs only in tumor cells, magnetic hyperthermia is gaining importance. Sadhukha et al. [36] developed EGFR-targeted superparamagnetic iron oxide NPs (SPIONs) and administered them to orthotopic lung tumor model via tracheal instillation and aerosol inhalation. EGFR-targeted SPIONs successfully concentrated and uniformly distributed in A549 and A549-luc (luciferase-transfected A549 for in vivo bioluminescent imaging) cell lines of orthotopic tumor models. Finally, they subjected the treated models to magnetic hyperthermia and observed that there was a significant inhibition of in vivo tumor growth with time (by monitoring the lung tumor bioluminescence as shown in Fig. 4.7).

MNPs are also often used as superior CAs in magnetic resonance imaging (MRI) over the most commonly clinically tried gadolinium-based CAs. Although MRI is one of the best imaging tools for biological systems especially for lung cancer cells, use of labeled nanomaterials generally yield poor results in MRI. Both low lung density as compared to other soft tissues and extensive air–tissue interfaces reduce MRI signals [37]. Generally, MNPs conjugated with active targeting ligands tend to concentrate in tumors and boost contrast of the affected tissues or organs in comparison with other body parts, through the acceleration of water proton relaxation in those areas. In a particular type of tissue, a group of water protons exhibit similar magnetic moments, i.e., they are in resonance. Now that all the water protons get excited when placed in a homogeneous magnetic field and their spins flip all together. When the field is removed, they collectively flip back to their original state and generate a radio frequency signal. This process of relaxation produces a signal which is captured through currents induced over specific arrangements of pickup coils. Finally, the whole relaxation process is

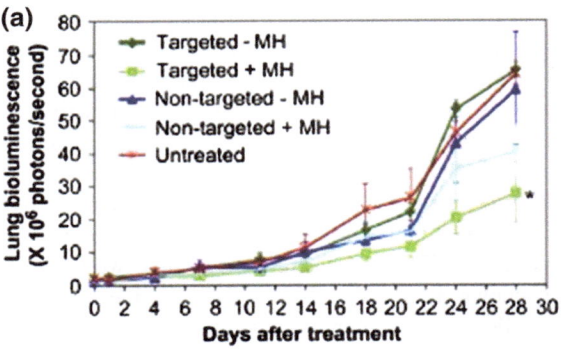

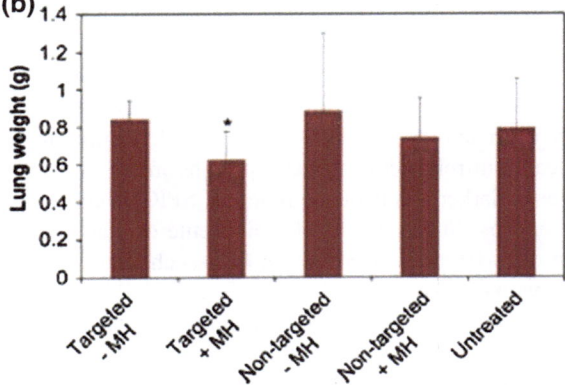

Fig. 4.7 Effect of targeted magnetic hyperthermia on lung tumor growth. Orthotopic lung tumor-bearing mice were allowed to inhale targeted or non-targeted SPIONs. After 1 week, 6 animals from each group were subjected to magnetic hyperthermia (MH) for 30 min. **a** Lung tumor bioluminescence was monitored over a period of 1 month. Data shown as mean ± SD ($n = 6$; *$P < 0.05$ compared to saline-treated and unheated controls). **b** Lungs were collected at the end of the efficacy study (1 month after magnetic hyperthermia) and weighed. Data shown as mean ± SD ($n = 6$; *$P < 0.05$ compared to unheated control). "Reprinted (adapted) with permission from Sadhukha et al. [36]. Copyright (2013) Elsevier"

tracked on a computer screen in the form of a temporal or spatial 2D or 3D images of the desired tissue. Relaxation signals are of the form:

$$m_z = m\left(1 - e^{-t/T_1}\right) \quad \text{and} \quad m_{xy} = m(\sin \omega_0 t + \Phi)e^{-t/T_2}$$

where T_1 and T_2 are longitudinal (spin–lattice) and transverse (spin–spin) relaxation times, respectively, and Φ is a phase constant. The longitudinal relaxation reflects a loss in the form of heat from the system to its surrounding lattice and is primarily a measure of the dipolar coupling of the proton moments to their surroundings. The relaxation in the transverse direction is due to loss of phase coherence in the processing protons owing to their magnetic interactions with each other and fluctuating field in the tissues. CAs are introduced to tag the target cells which change either of the relaxation times (T_1 and T_2) of the surrounding tagged hydrogen atoms to a measurable extent in the form of signal contrast intensity that distinguishes between the charged (malignant) and normal tissues. However, they are not magnetic in the absence of external magnetic field and thus can be utilized in vivo safely, eliminating any risk of harmful radiation [38]. Actually, CAs are believed to be more concentrated in the malignant parts owing to larger water proton relaxation in those regions [39]. Performance of CAs, i.e., their ability to influence the relaxation times, depends on the value of the square of their saturated magnetic moments (M_s). The larger the value of M_s, the better the performance of MRI CAs. There are basically two types of MRI CAs: one is T_1 MRI CAs (also known as positive CAs as they provide brighter images) such as gadolinium(III) complexes with a high longitudinal relaxivity of 3–5 s^{-1} mM^{-1} owing to the seven unpaired electrons in its 4f orbital, whereby yielding the largest electron spin magnetic moments combined with slow spin relaxation of the S state electrons. This closely matches the spin relaxation state of water protons and thus by resonance effect provides a higher signal contrast. On the other hand, another is T_2 MRI CAs (also known as negative CAs as they provide darker images) such as SPIONs with a very large transverse relaxivity of 100–200 s^{-1} mM^{-1}. Tumor-directing SPIONs can change the nuclear spin relaxation of water protons in the affected areas and cause the region of interest darker. Available commercial SPIONs consist of a core of either monocrystalline or cross-linked iron oxide (magnetite or maghemite) having a diameter in the range of 5–10 nm, coated with a polysaccharide usually dextran or other synthetic biocompatible polymers such as PEG and PVP to ensure longtime blood circulation [40]. However, such polymeric capping is attached to the iron oxide core only through non-covalent interactions and is likely to get detached during circulation in the bloodstream, thus unsuitable for cancer imaging. Lee et al. [40] reported that thermally cross-linked antibiofouling polymer like poly(TMSMA-r-PEGMA)-coated SPIONs is very effective as T_2 MRI CAs. Here, polymeric block of 3-(trimethoxysilyl) propyl methacrylate (TMSMA) helped in surface anchoring with SPIONs owing to the presence of silane groups and that of poly(ethylene glycol) methyl ether methacrylate (PEGMA) provided biocompatibility and physiological stability owing to the presence of protein-resistant moieties. Such systems of antibiofouling polymer-coated TCL-SPIONs being stable and antifouling in nature (protein resistant) easily accumulated in the target cells by EPR effect even in the absence of any active targeting ligands. Thus, they reached any remote areas including lungs and liver from where generally NPs tend to excrete out of the body. Yu et al. [41] described that when antibiofouling polymer-coated carboxyl TCL-SPIONs loaded with DOX were injected into the back of LLC grafted mice, affected areas appeared dark in MR images as shown in Fig. 4.8. Even the respective fluorescent images were bright only in the tumor zone at the back of the mice in comparison with other organs.

Yu et al. also showed that the system of Dox@TCL-SPION successfully delivered anticancer drugs to the tumor very specifically as the drug was attached to the MNPs only which selectively get internalized in the tumors. In an attempt to reduce side effects due to non-uniform and non-specific distribution of chemotherapeutic drugs and also to minimize the dosage of administration of the same, MNPs have emerged as a potential drug vehicle for specific tumor targeting. Often, in other systems, the

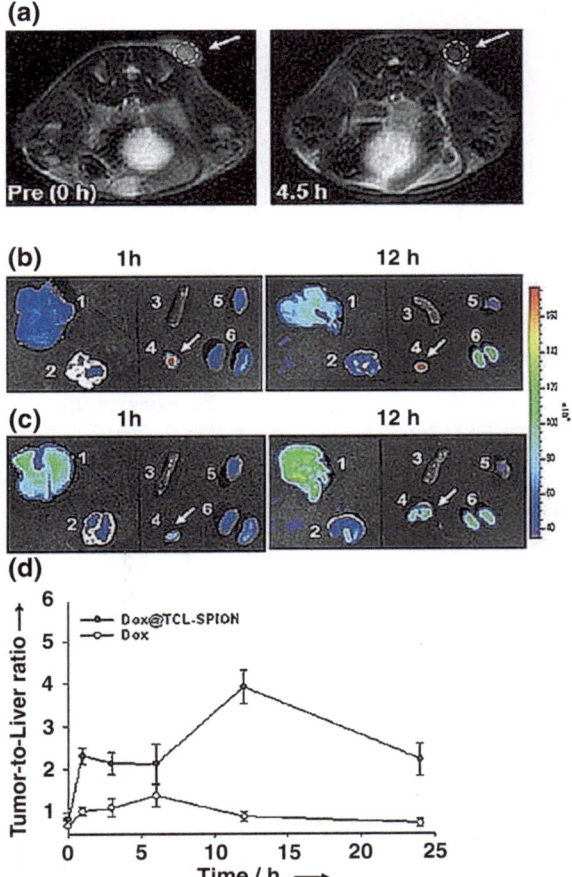

Fig. 4.8 a T$_2$-weighted fast spin-echo images (time of repetition/time of echo: 4,200 ms/102 ms) taken at 0 and 4.5 h after injection of Dox@TCL-SPION at the level of LLC tumor on the right back of the mouse. The *dashed circle* with *white arrow* indicates the allograft tumor region. **b, c** Optical fluorescence images of major organs and allograft tumors: 1 liver; 2 lung; 3 spleen; 4 tumor; 5 heart; 6 kidney. Images were taken after intravenous injection of **b** Dox@TCL-SPION (equivalent to 4 mg of Dox) and **c** free Dox (4 mg) into tumor-bearing mice (*n* = 3); mice were euthanized after 1 and 12 h. **d** The ratio of fluorescence intensities of tumor to liver for Dox@TCL-SPION (*closed circles*) and free Dox (*open circles*) as a function of time. "Reprinted (adapted) with permission from Yu et al. [41]. Copyright (2008) John Wiley and Sons"

principles guiding the magnetic therapy are similar to those followed in magnetic separation. A conjugate of a cytotoxic drug (say DOX) and a biocompatible MNP in the form of ferrofluid often targets the specific tumor by an external magnetic field (active targeting) (Fig. 4.9). Finally, the drug sequestered from the MNP surface to the specific site via enzymatic activity or in response to any physiological changes in the tumor zones such as pH, temperature, and concentration.

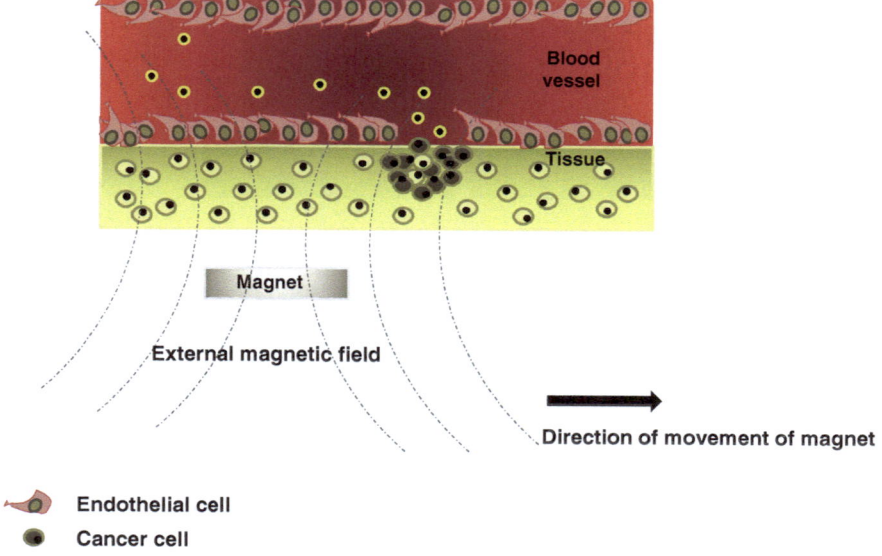

Fig. 4.9 Active targeting of MNP–drug conjugate to cancer cells under external magnetic field

MNPs have also circumvented the problems of chemotherapeutic drug resistance in the treatment of recurring cancers such as lung cancer to a significant extent. A team of researchers in a preclinical trial showed that when MNPs loaded with cisplatin were administered to a cisplatin-resistant lung-tumor-xenografted model, there was reduction in localized drug-resistant proteins in the tumor microenvironment and exhibited high cellular toxicity [42].

4.3 Quantum Dots

Multiplexing (simultaneous detection of multiple signals) of tumor markers in biological samples has opened up a new promising field in the early detection of cancer, differentiation of damaged cells from the healthy benign cells, evaluation of the stage of the disease, response of the tumors to therapy, and prediction of recurrence. Currently employed diagnostic techniques such as medical imaging, tissue biopsy, and bioanalytical survey of body fluids by ELISA lack sensitivity and specificity to detect lung cancer at an early stage. Also, the techniques are tedious, elaborate, and expensive and do not have any multiplexing capability. On the contrary, QDs-based detection is rapid, easy, and economically viable for lung cancer screening. QDs are extensively used for in vivo biomolecular and cellular imaging of cancer cells by monitoring the analytes of cancer markers, present even at very low concentration. QDs exhibit high photo- and chemical stability, are resistant to photobleaching under high illumination intensity, possess size-tunable long-term multicolor auto-fluorescence property, produce broad overlapping absorption spectra combined with very narrow symmetric emission spectra (full width at half maximum ~25–40 nm) [43], and display superior signal brightness due to high molar extinction coefficients (~10–100 × organic dyes [43]). QDs are basically semiconductor-based nanocrystals/particles having size in the range of 1–10 nm with unique photochemical and photophysical properties unlike common organic dyes and fluorescent proteins [44]. QDs are composed of atoms from group II and VI elements [e.g., cadmium selenide (CdSe), CdTe] or group III and V elements (e.g., InP, InAs) [45]. Their physical dimensions are smaller than the exciton Bohr radius that leads to the quantum confinement effect (i.e., charge carriers are confined within the nanoscale dimensions), which is responsible for their unique optical and electronic properties [45, 46]. QDs emit light of different colors (in the range of UV to IR) depending upon their sizes, when excited at appropriate wavelengths. In general, a smaller particle emits shorter wavelength of light and vice versa. Such biological levels exhibit redshift of emission peaks with respect to absorption spectra (Stokes' shift) and particle size [47]. One of the striking features of QDs is that multicolor QDs can be simultaneously excited by a single light source at a wavelength below the wavelength of fluorescence (Fig. 4.10) [45]. Long-term photostability of QDs makes them ideal for investigating various cellular dynamics over time. Often, the derivatives of QDs are linked to immunoglobulins to develop a fluorescent imaging system for the detection of specific antigens bound to the damaged cells. QDs may also be covalently bonded to a panel of molecules (antibodies, peptides, nucleic acids, and other ligands) due to their high aspect ratio, making them ideal for designing complex multifunctional nanostructures suited for fluorescence probing applications in targeting malignant tumors with high specificity [45]. Another importance of QDs is that they can hold multitude of CAs and develop a multicomponent diagnostic system to present the metabolic state of different cells. Owing to the unique nanodimensional feature of QDs, such multicomponent diagnostic systems are very sensitive to detect any abnormalities within the body at a very early stage. This is a very accepting field of diagnosis as compared to the present available methods, whereby a single molecule of an MRI CA is bound to a monoclonal antibody and then targeted to the cancer cells and sensed. This requires thousands of such conjugate carriers to bind to the cancer cells to generate strong enough signals in order to be detected via an MRI. However, if the

Fig. 4.10 Emission colors of ZnS-capped CdSe QDs excited with a near UV lamp (*Credit* Image courtesy by http://ffden-2.phys.uaf.edu/211_fall2004.web.dir/jason_turnquist/main.htm, Access date 28 Apr 2014)

Fig. 4.11 Various assembled components in quantum dots

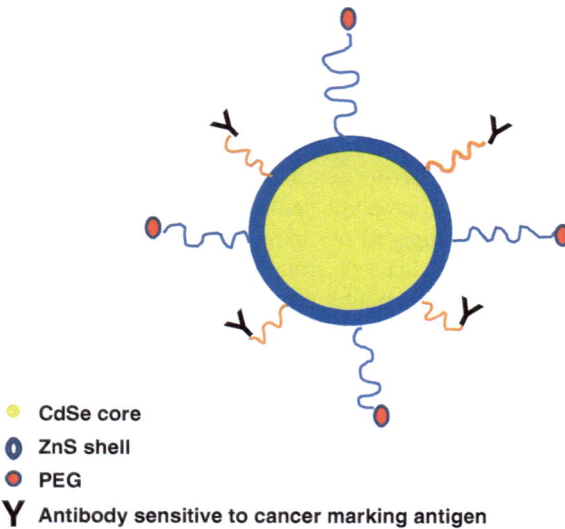

● CdSe core
◖ ZnS shell
● PEG
Y Antibody sensitive to cancer marking antigen

same cancer-sensing monoclonal antibody is attached to a NP containing thousands of MRI CAs, then this single conjugate is enough to create a sharp peak when bound to cancer cells. Obviously, the latter method of detection of anomaly in biological system is very sensitive and highly efficient due to self-fluorescence property of QDs.

The most commonly used QD fluorophores in biology are the semiconductor colloidal nanocrystals whose inner core is made up of CdSe and the outer shell is made up of zinc sulfide (ZnS) as depicted in Fig. 4.11. ZnS shell passivates the core CdSe, protects it from corrosion, prevents leaching of the inner components to the surrounding solution, and improves the photoluminescence signal. However, there lies a potential life threat from CdSe-based QDs due to the inherent toxic nature of Cd which may be released in the ionic form, from the inner core when the targeted cells already concentrated with QDs are exposed to the UV light [45, 48]. Moreover, naked QDs are water insoluble which retard their circulation in the human circulatory system. Keeping these facts in mind, QDs are generally encapsulated in the biocompatible polymers (PEG, PVA, PEGylated dihydrolipoic acid, dendrimers, multidentate phosphine polymers [49]), and amphiphilic polymers (poly(maleic anhydride alt-1-tetradecene, triblock copolymer, alkyl-modified poly(acrylic acid) [49]) to render them soluble in cellular environment (mostly aqueous based), resistant to agglomeration, and resistant to degradation in biological environment, thereby preventing them from becoming threat to life. QD bioconjugates may be introduced into the human body by both active and passive targeting mechanisms although the latter is slow and less efficient [45, 50]. QDs may identify various potential biomarkers of lung cancer which include specific proteins, DNA, mRNA sequences, and circulating tumors. Dual-color QDs have been used successfully in performing a simultaneous dual-protein immunoassay of lung cancer-based biomarkers carcinoembryonic antigen (CEA) and neuron-specific enolase (NSE) in a single sample [51]. Akerman and his coworkers conjugated QDs with three different lung cancer cells targeting peptides such as CGFECVRQCPERC sequence (GFE), KDEPQRRSARLSAKPAPPKPEPKPKKAPAKK (F3), and CGNKRTRGC (LyP-1) [52]. Various blood vessels express molecular markers that distinguish the vasculature of individual organs, tissues, and tumors. Thus, they used specifically GFE to target membrane dipeptidase on the endothelial cells in lung blood vessels, F3 to bind blood vessels and tumor cells in various tumors, and LyP-1 to recognize lymphatic vessels and tumor cells in certain tumors. In a typical process, they synthesized tri-n-octylphosphine oxide-coated ZnS-capped CdSe QDs, followed by coating of mercaptoacetic acid. Finally, they conjugated the mercaptoacetic acid-coated QDs to thiolated amine-terminated PEG (the outer polymeric coating over the QDs prevents non-selective accumulation of QDs in the reticuloendothelial tissues) and attached the whole complex to respective thiolated peptide ligands by simple acid–base reaction. They found that lung homing green GFE QDs selectively decorated the surface of lung endothelial (LE) cells in a model xenografted with LE, brain endothelial, and human breast carcinoma MDA-MD-435 cells. As LE cells express membrane dipeptidase (the receptor for GFE peptide), the GFE QDs only targeted the LE cells but failed to target brain endothelial cells. QDs may also execute intraoperative sentinel lymph nodes (SLNs) mapping by NIR fluorescence property, emitting light at 850 nm. SLNs are the first lymph nodes or group of nodes invaded by the metastasizing cells from a primary NSCLC tumor of lungs [53, 54]. This is basically deep tissue imaging where NIR/IR emitting QDs are essential in increasing the tumor imaging sensitivity as the Rayleigh scattering decreases with increasing wavelength and the major absorption peaks of blood and water do not interfere in this region [45, 55, 56]. Moreover, the heat generated from the NIR-based QDs also helps to destroy the surrounding roaming cancer cells [45]. This type of QD technology provides the exact image of the lymph node to the surgeon in executing successful operation. DNA methylation assay using methylation-specific QDs fluorescence resonance energy transfer (MS-qFRET) for the early detection of lung cancer is a rapidly developing sensitive methodology in an attempt to preserve life. Inactivation of tumor suppressor gene is quoted as one of the potential reasons behind any forms of cancer. A biochemical change called DNA methylation occurs when a methyl group attaches itself to cytosine, one of the four nucleotides or base building blocks of DNA. When DNA methylation occurs at critical gene locations (carbon 5 position of cytosine in palindromic CpG dinucleotides), it can halt the release of proteins that suppress tumors and thus encourage development of cancers. Hypermethylation of CpG islands in the promoter regions of PYCARD genes with the suppression of ASC/TMS1 tumor suppressor protein is a common phenomenon in lung cancer especially at the later stage of the disease, as demonstrated by the analysis of the methylation status of over 40 genes from lung cancer tumors, cell lines, patient sputum, and/or serum [57, 58]. Also, CDKN2A gets methylated at an early stage of lung cancer. In this context, identification of this attachment of methyl groups to DNA strands of the patient's sample may prove to be very helpful in lung cancer therapy. In a typical process, Bailey et al. [59] converted all the normal segments of DNA that lack methyl groups into uracil leaving the already methylated cytosine unaffected through a chemical process called bisulfite conversion. Then, they performed polymerase chain reaction (PCR) amplification of the bisulfite-treated DNA wherein the former primer was biotinylated and the reverse primer was labeled with an organic fluorophore. Here, only a very few DNA strands with methyl groups on cytosine (abnormal gene) were amplified selectively so that their abnormal sequence was detectable. Finally, they mixed the PCR-amplified DNA strands with streptavidin-conjugated QDs. Streptavidin ligands being highly affectionate to biotin moieties on the customized DNA strands, streptavidin-conjugated QDs readily labeled the PCR-amplified DNA strands. Approximately 60 of the targeted DNA strands struck themselves to a single QD, developing an octopus-like structure. Now that owing to the streptavidin–biotin binding, DNA arms bearing QDs (donors) and fluorophores (acceptors) came to such a close proximity that fluorescence resonance energy transfer (FRET) occurred when an UV or a blue light laser shines on a QD. During the transfer of energy of the light from QD to the neighboring fluorescent dye molecule, a fluorescent glow was emitted and indicated the presence of the abnormal DNA strands almost instantaneously as shown in Fig. 4.12.

Apart from the quality analysis, quantitative analysis is also another platform for cancer diagnosis. Intensity of the fluorescence emissions gives an idea of the stage of cancer. Obviously, higher emission means higher number of abnormal DNA strands and thus correlated with higher cancer risk. Frequent and routine analysis through this test also helps doctors to track whether a particular treatment is working or not. The test also identifies specific genetic markers associated with particular cancer type (say lung cancer). In brief, DNA methylation test is an upcoming cancer diagnosis tool which may soon lead the clinical trials.

Fig. 4.12 In this illustration by Yi Zhang, QDs are depicted as *gold spheres* that attract DNA strands linked to cancer risks. When the QDs were exposed to certain types of light, they transferred the energy to fluorescent molecules, shown as *pink* globes that emitted a glow. This enabled researchers to detect and count the DNA strands linked to cancer. (*Credit* Image courtesy of Johns Hopkins University, http://www.understandingnano.com/quantum-dots-dna-test-early-cancer-detection.html, Access date 04 May 2014)

4.4   Silica Nanoparticles

Silica nanoparticles (SiNPs) form an important class of ceramic NPs, especially for cancer therapy. They are inexpensive, easy to prepare, and biocompatible [60]. Surface functionalization of SiNPs also makes them unparallel candidates as solid support for immobilization of biological entities such as enzymes, proteins, and DNA and makes them suitable for sensing, cellular imaging, drug delivery, and gene transfection [61, 62]. Mesoporous silica is another developing field. They have high surface areas and porous interiors. Often, the drugs for cancer treatment are hydrophobic, and thus, they suffer from irregular distribution of the same in the hydrophilic bloodstream to the tumor cells, often with inevitable loss of drug efficiency through excretion out of the target locations. Mesoporous silica facilitates encapsulation of hydrophobic chemotherapeutic drugs (say camptothecin derivative) and delivery of the same through the circulatory system to the specific sites without concomitant loss of its efficiency. Zhang et al. [63] synthesized a novel molecular imaging agent of miRNA-targeting oligonucleotide-functionalized nanoshells comprising of silver shells and silica cores encapsulating fluorescent $Ru(bpy_3)^{2+}$ complexes to detect miRNA-486 at a very early stage with high precision, a potential biomarker in lung cancer-positive cell lines (H460 and H1944) by fluorescence in situ hybridization (FISH) technique. They used miRNA

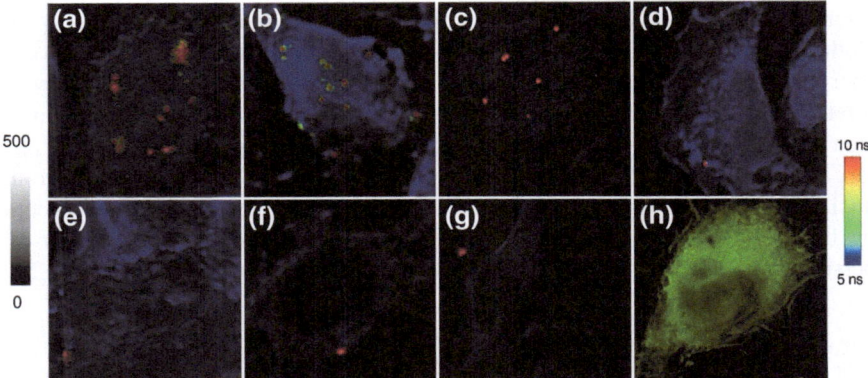

Fig. 4.13 Representative fluorescence images from the single cells that were incubated by the metal nanoshell with the single-stranded probe oligonucleotide conjugation (*top panel*) **a** H460, **b** H1944, **c** MDA-MB-231, and **d** A549 or incubated by the metal nanoshell without the probe oligonucleotide conjugation (bottom panel): **e** H460, **f** H1944, and **g** MDA-MB-231. Images **a–g** have the emission intensity bar from 0 to 500 and the lifetime bar from 5 to 10 ns. Image **h** represents the cell image of H1944 hybridized with Cy3-labeled oligonucleotide probe. Image **h** has the emission intensity bar from 0 to 200 and the lifetime bar from 1 to 4 ns. The scales of **a** and **e** diagrams are 20 × 20 μm, and others are 50 × 50 μm. The resolutions of diagrams are an integration of 0.6 ms/pixel. "Reprinted (adapted) with permission from Zhang et al. [63]. Copyright (2010) American Chemical Society."

non-expressing cancer cells of A549 as control. They obtained fluorescent cell images of the various miRNA (positive and negative)-expressing cancer cell lines incubated with fluorescent nanoshell probe on confocal microscope. They found that owing to the stronger intensity and longer lifetime of the nanoshell imaging probe, the emission signals from the metal nanocells could be isolated distinctly from the cellular autofluorescence in the images as shown in Fig. 4.13 and also they could count the emission spots at the single miRNA level. They verified the selective hybridization of the imaging probe with the miRNAs by performing control experiments in which they did not use miRNA-targeting oligonucleotide and observed reduced emission signals.

However, use of SiNPs (amorphous form) till date is very limited. SiNPs induce toxicity to healthy cells and cause inflammation of target organs which often lead to apoptosis of normal cells. Hence, concentration of SiNPs is never exceeded beyond 0.1 mg/mL even in in vitro tests. Such a low concentration of SiNPs fails to bring out an effective therapy against cancer as we know many of them get cleared by the RES.

4.5 Lanthanide Nanoparticles

Lanthanides or rare earth metals are elements with atomic number 57 (lanthanum) to 71 (lutetium) including various stable isotopes. NPs of lanthanides (LNPs) are often used to tag cells or biological macromolecules for biodiagnostic assay owing to their inherent photoluminescent properties with the ability to upconvert low-energy light (NIR) to high-energy light (UV–Vis) (Fig. 4.14) [64]. Thus, they are often called upconverting NPs (UCNPs). One of the main reasons for upconversion is the generation of more than one absorbed photon per single emitted photon. The paramagnetic (4f) electrons of LNPs are compactly localized close to the nucleus, and thus, their magnetic and fluorescent properties are mostly unaffected by ligand effects of biocompatible coatings such as hydrophilic and water-dispersible polyethyleneimine [39, 65]. As the UCNPs are photoexcitable in the NIR

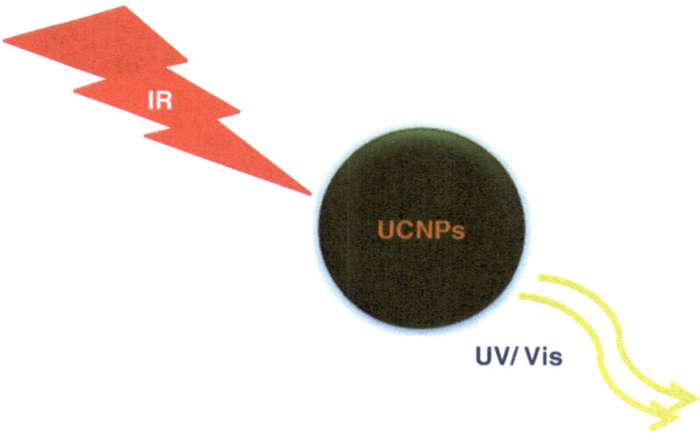

Fig. 4.14 Upconversion of light by LNPs

(biological window), they limit any background cellular absorption and autofluo-rescence. Again, the emission bands by UCNPs are very narrow, enable easy sepa-ration, and enhanced selectivity of imaging assay. Such UCNPs are also reported to possess size-independent emission. However, all lanthanides do not exhibit upconversion luminescence. Tripositive erbium (Er^{3+}), thulium (Tm^{3+}), and hol-mium (Ho^{3+}) have shown successful upconversion luminescence owing to their structure. Often, the upconversion efficiency is improved by synergistic action of ytterbium ions (Yb^{3+}). Generally, for cellular imaging, single-phase lanthanide oxide NPs are considered owing to their decreased d value, narrow fluorescent bandwidth unlike organic dyes, very high photostability, and non-toxicity in bio-logical systems.

One of the advantages offered by LNPs over QDs is reported to be its exci-tation in the NIR region which allows harmless and deep tissue penetration and thereby better image resolution. This is probably due to low degree of Rayleigh scattering while using NIR light over UV light. Another interesting feature is that in Ln^{3+} ions, excitation occurs via real electronic states of defined and relatively long lifetimes. Thus, weak and cheap laser diodes may be used to excite LNPs. Often, upconverting or downconverting fluorescent Ln(III) ions are doped with paramagnetic MRI CA NPs of Dy_2O_3, Gd_2O_3, etc., in an attempt to make con-ventional MRI CAs as MRI-FI agent. Actually, multiplex imaging tools provide complementary information on a disease [39]. Even the paramagnetic LNPs may be further functionalized with target-specific anticancer drug molecules for can-cer treatment along with imaging. This system was commercialized for molecular diagnosis of lung cancer in a R&D project of MKE/KEIT (Grant No: 10040393).

References

1. Salem H, Eid K, Sharaf M (2011) Formulation and evaluation of silver nanoparticles as anti-bacterial and antifungal agents with a minimal cytotoxic effect. Int J Drug Deliv
2. Link S, El-Sayed MA (2003) Optical properties and ultrafast dynamics of metallic nanocrys-tals. Annu Rev Phys Chem 54:331–366
3. Kelly KL, Coronado E, Zhao LL, Schatz GC (2002) The optical properties of metal nanopar-ticles: the influence of size, shape, and dielectric environment. J Phys Chem B 107:668–677
4. Boisselier E, Astruc D (2009) Gold nanoparticles in nanomedicine: preparations, imaging, diagnostics, therapies and toxicity. Chem Soc Rev 38:1759–1782
5. Cherukuri P, Glazer ES, Curley SA (2010) Targeted hyperthermia using metal nanoparticles. Adv Drug Deliv Rev 62:339–345
6. Ng VWK, Berti R, Lesage F, Kakkar A (2013) Gold: a versatile tool for in vivo imaging. J Mater Chem B 1:9–25
7. Choi YE, Kwak JW, Park JW (2010) Nanotechnology for early cancer detection. Sensors (basel) 10:428–455
8. Bhattacharya R, Mukherjee P (2008) Biological properties of "naked" metal nanoparticles. Adv Drug Deliv Rev 60:1289–1306
9. Bhattacharyya S, Kudgus R, Bhattacharya R, Mukherjee P (2011) Inorganic nanoparticles in cancer therapy. Pharm Res 28:237–259
10. Patra CR, Bhattacharya R, Mukhopadhyay D, Mukherjee P (2008) Application of gold nano-particles for targeted therapy in cancer. J Biomed Nanotechnol 4:99–132

11. Haiduc I, Silvestru C (1989) Rhodium, iridium, copper and gold antitumor organometallic compounds (review). Vivo 3:285–294

12. Phillips M, Cataneo RN, Cummin ARC, Gagliardi AJ, Gleeson K, Greenberg J, Maxfield RA, Rom WN (2003) Detection of lung cancer with volatile markers in the breath. Chest 123:2115–2123

13. O'Neill HJ, Gordon SM, O'Neill MH, Gibbons RD, Szidon JP (1988) A computerized classification technique for screening for the presence of breath biomarkers in lung cancer. Clin Chem 34:1613–1618

14. Hao Y, Liang X, Mingfu C, Xing C, Ping W, Jiwei J, Yuelin W (2003) Detection volatile organic compounds in breath as markers of lung cancer using a novel electronic nose. Proc IEEE Sens 2:1333–1337

15. Phillips M, Gleeson K, Hughes JMB, Greenberg J, Cataneo RN, Baker L, McVay WP (1999) Volatile organic compounds in breath as markers of lung cancer: a cross-sectional study. Lancet 353:1930–1933

16. Peng G, Tisch U, Adams O, Hakim M, Shehada N, Broza YY, Billan S, Abdah-Bortnyak R, Kuten A, Haick H (2009) Diagnosing lung cancer in exhaled breath using gold nanoparticles. Nat Nanotechnol 4:669–673

17. Phillips M, Cataneo RN, Greenberg J, Grodman R, Gunawardena R, Naidu A (2003) Effect of oxygen on breath markers of oxidative stress. Eur Respir J 21:48–51

18. Mitsui T, Kondo T (2003) Inadequacy of theoretical basis of breath methylated alkane contour for assessing oxidative stress. Clin Chim Acta 333:91

19. Smith D, Wang T, Sulé-Suso J, Španěl P, Haj AE (2003) Quantification of acetaldehyde released by lung cancer cells in vitro using selected ion flow tube mass spectrometry. Rapid Commun Mass Spectrom 17:845–850

20. Kamat PC, Roller CB, Namjou K, Jeffers JD, Faramarzalian A, Salas R, McCann PJ (2007) Measurement of acetaldehyde in exhaled breath using a laser absorption spectrometer. Appl Opt 46:3969–3975

21. Giubileo G (2002) Medical diagnostics by laser-based analysis of exhaled breath. Proc SPIE 318–325

22. Di Natale C, Macagnano A, Martinelli E, Paolesse R, D'Arcangelo G, Roscioni C, Finazzi-Agro A, D'Amico A (2003) Lung cancer identification by the analysis of breath by means of an array of non-selective gas sensors. Biosens Bioelectron 18:1209–1218

23. Mukherjee P, Bhattacharya R, Wang P, Wang L, Basu S, Nagy JA, Atala A, Mukhopadhyay D, Soker S (2005) Antiangiogenic properties of gold nanoparticles. Clin Cancer Res 11:3530–3534

24. Sandler A, Gray R, Perry MC, Brahmer J, Schiller JH, Dowlati A, Lilenbaum R, Johnson DH (2006) Paclitaxel-carboplatin alone or with bevacizumab for non-small-cell lung cancer. Engl J Med 355:2542–2550

25. Conde J, Doria G, Baptista P (2012) Noble metal nanoparticles applications in cancer. J Drug Deliv 12

26. Cheng FY, Chen CT, Yeh CS (2009) Comparative efficiencies of photothermal destruction of malignant cells using antibody-coated silica@Au nanoshells, hollow Au/Ag nanospheres and Au nanorods. Nanotechnology 20:425104

27. Kim DM, Noh HB, Park DS, Ryu SH, Koo JS, Shim YB (2009) Immunosensors for detection of Annexin II and MUC5AC for early diagnosis of lung cancer. Biosens Bioelectron 25:456–462

28. Rahman MA, Won MS, Shim YB (2005) The potential use of hydrazine as an alternative to peroxidase in a biosensor: comparison between hydrazine and HRP-based glucose sensors. Biosens Bioelectron 21:257–265

29. Shi X, Wang S, Meshinchi S, Van Antwerp ME, Bi X, Lee I, Baker JR (2007) Dendrimer-entrapped gold nanoparticles as a platform for cancer-cell targeting and imaging. Small 3:1245–1252

30. Arvizo R, Bhattacharya R, Mukherjee P (2010) Gold nanoparticles: opportunities and challenges in nanomedicine. Expert Opin Drug Deliv 7:753–763

31. Chen YH, Tsai CY, Huang PY, Chang MY, Cheng PC, Chou CH, Chen DH, Wang CR, Shiau AL, Wu CL (2007) Methotrexate conjugated to gold nanoparticles inhibits tumor growth in a syngeneic lung tumor model. Mol Pharm 4:713–722

32. Goya GF, Grazu V, Ibarra MR (2008) Magnetic nanoparticles for cancer therapy. Curr. Nanosci. 4:1–16

33. Pankhurst QA, Thanh NTK, Jones SK, Dobson J (2009) Progress in applications of magnetic nanoparticles in biomedicine. J Phys D: Appl 42:224001

34. Fuentes M, Mateo C, Guisan JM, Fernandez-Lafuente R (2005) Preparation of inert magnetic nano-particles for the directed immobilization of antibodies. Biosens Bioelectron 20:1380–1387

35. Wang Y, Zhang Y, Du Z, Wu M, Zhang G (2012) Detection of micrometastases in lung cancer with magnetic nanoparticles and quantum dots. Int J Nanomed 7:2315–2324

36. Sadhukha T, Wiedmann TS, Panyam J (2013) Inhalable magnetic nanoparticles for targeted hyperthermia in lung cancer therapy. Biomaterials 34:5163–5171

37. Garbow J, Orellana C, Neil J (2005) Magnetic resonance imaging. In: Schuster DP, Blackwell TS (eds) Molecular imaging of the lungs, vol 203. CRC Press, Boca Raton, pp 113–133

38. Saboktakin MR, Maharramov A, Ramazanov MA (2009) Synthesis and characterization of superparamagnetic nanoparticles coated with carboxymethyl starch (CMS) for magnetic resonance imaging technique. Carbohydr Polym 78:292–295

39. Xu W, Kattel K, Park JY, Chang Y, Kim TJ, Lee GH (2012) Paramagnetic nanoparticle T1 and T2 MRI contrast agents. Phys Chem Chem Phys 14:12687–12700

40. Lee H, Lee E, Kim DK, Jang NK, Jeong YY, Jon S (2006) Antibiofouling polymer-coated superparamagnetic iron oxide nanoparticles as potential magnetic resonance contrast agents for in vivo cancer imaging. J Am Chem Soc 128:7383–7389

41. Yu MK, Jeong YY, Park J, Park S, Kim JW, Min JJ, Kim K, Jon S (2008) Drug-loaded superparamagnetic iron oxide nanoparticles for combined cancer imaging and therapy in vivo. Angew Chem Int Ed 47:5362–5365

42. Sun W, Fang N, Trewyn B, Tsunoda M, Slowing I, Lin VY, Yeung E (2008) Endocytosis of a single mesoporous silica nanoparticle into a human lung cancer cell observed by differential interference contrast microscopy. Anal Bioanal Chem 391:2119–2125

43. Medintz IL, Uyeda HT, Goldman ER, Mattoussi H (2005) Quantum dot bioconjugates for imaging, labelling and sensing. Nat Mater 4:435–446

44. Smith AM, Gao X, Nie S (2004) Quantum dot nanocrystals for in vivo molecular and cellular imaging. Photochem Photobiol 80:377–385

45. Vashist S, Tewari R, Bajpai RP, Bharadwaj LM, Raiteri R (2007) Review of quantum dot technologies for cancer detection and treatment. J Nanotechnol

46. Chan WCW, Maxwell DJ, Gao X, Bailey RE, Han M, Nie S (2002) Luminescent quantum dots for multiplexed biological detection and imaging. Curr Opin Biotechnol 13:40–46

47. Gao X, Cui Y, Levenson RM, Chung LWK, Nie S (2004) In vivo cancer targeting and imaging with semiconductor quantum dots. Nat Biotechnol 22:969–976

48. Derfus AM, Chan WCW, Bhatia SN (2003) Probing the cytotoxicity of semiconductor quantum dots. Nano Lett 4:11–18

49. Wang Y, Chen L (2011) Quantum dots, lighting up the research and development of nanomedicine. Nanomed Nanotech Biol Med 7:385–402

50. Vicent MJ, Duncan R (2006) Polymer conjugates: nanosized medicines for treating cancer. Trends Biotechnol 24:39–47

51. Li H, Cao Z, Zhang Y, Lau C, Lu J (2011) Simultaneous detection of two lung cancer biomarkers using dual-color fluorescence quantum dots. Analyst 136:1399–1405

52. Åkerman ME, Chan WCW, Laakkonen P, Bhatia SN, Ruoslahti E (2002) Nanocrystal targeting in vivo. Proc Natl Acad Sci USA 99:12617–12621

53. Luo G, Long J, Zhang B, Liu C, Ji S, Xu J, Yu X, Ni Q (2012) Quantum dots in cancer therapy. Expert Opin Drug Deliv 9:47–58

54. Schmidt F, Woltering E, Webb W, Garcia O, Cohen J, Rozans M (2002) Sentinel nodal assessment in patients with carcinoma of the lung. Ann Thorac Surg 74:870–874

55. Ntziachristos V, Bremer C, Weissleder R (2003) Fluorescence imaging with near-infrared light: new technological advances that enable in vivo molecular imaging. Eur Radiol 13:195–208
56. Cheong WF, Prahl SA, Welch AJ (1990) A review of the optical properties of biological tissues. IEEE J Quantum Electronics 26:2166–2185
57. Tsou JA, Hagen JA, Carpenter CL, Laird IA (2002) Offringa, DNA methylation analysis: a powerful new tool for lung cancer diagnosis. Oncogene 21:5450–5461
58. Machida EO, Brock MV, Hooker CM, Nakayama J, Ishida A, Amano J, Picchi MA, Belinsky SA, Herman JG, Taniguchi SI, Baylin SB (2006) Hypermethylation of ASC/TMS1 is a sputum marker for late-stage lung cancer. Cancer Res 66:6210–6218
59. Bailey VJ, Easwaran H, Zhang Y, Griffiths E, Belinsky SA, Herman JG, Baylin SB, Carraway HE, Wang T-H (2009) MS-qFRET: a quantum dot-based method for analysis of DNA methylation. Genome Res 19:1455–1461
60. Choi J, Zheng Q, Katz HE, Guilarte TR (2010) Silica-based nanoparticle uptake and cellular response by primary microglia. Environ Health Perspect 118:589–595
61. Feifel SC, Lisdat F (2011) Silica nanoparticles for the layer-by-layer assembly of fully electro-active cytochrome c multilayers. J Nanobiotechnol 9:59
62. Lu J, Liong M, Zink JI, Tamanoi F (2007) Mesoporous silica nanoparticles as a delivery system for hydrophobic anticancer drugs. Small 3:1341–1346
63. Zhang J, Fu Y, Mei Y, Jiang F, Lakowicz JR (2010) Fluorescent metal nanoshell probe to detect single miRNA in lung cancer cell. Anal Chem 82:4464–4471
64. Auzel F (2003) Upconversion and anti-stokes processes with f and d ions in solids. Chem Rev 104:139–174
65. Chatterjee DK, Rufaihah AJ, Zhang Y (2008) Upconversion fluorescence imaging of cells and small animals using lanthanide doped nanocrystals. Biomaterials 29:937–943

Chapter 5
Conclusion

5.1 Future Prospects of Nanoparticles for Lung Cancer Therapy

In the last few decades, nanotechnology has entered every phase of consumer products, especially in the design of medicines for the diagnosis and treatment of terminating illness like cancer. Undoubtedly, extensive research in nanobiomedicine will definitely produce safe, highly efficient, and highly sensitive medicines for oral and intravenous administrations and for early-stage cancer detection and monitoring schemes. These systems are already improving key issues like specific targeting, increase therapeutic efficacies, reduce side effects, and overcome drug resistance to certain extent. However, this is quite regretful that despite the development of many potential nanoparticle-based imaging probes and medicinal formulations for the diagnosis and treatment of lung cancer, hardly very few of them have been tried clinically and available in the market. Although synthesis of nanoparticles is very easy and environment friendly, they lack reproducibility, i.e., a particular synthetic scheme may not always produce nanoparticles of the same size and activity. Thus, broad particle size distribution, uncertainty in shape/size, and unpredictable surface chemistry of nanoparticles may lead to high risks when used in human body. Again there are also problems of undesirable immune reactions and rapid clearance from the body by the hepatic portal systems before any accumulation at the target locations, poor targeting, inability to infiltrate deep-lying tissues like those of lungs and development of high-resistant tumors against any anticancer formulations. Hence, FDA hardly approves any nanomedicine for cancer clinical trials. Further researches have to be done to improve synthesis methods for various nanoparticles so that reproducibility in physicochemical properties of nanoparticles is obtained for their safe use in human bodies. Often introduction of surface charges circumvents the problem of cell targeting. Particle–particle interaction in biological systems and aggregation tendencies of various nanoparticles highly depend on their zeta potential. Cellular membranes are generally negatively charged owing to the presence of nucleic acids. Hence, any positively charged nanoparticles may easily target tumor cells. In fact, nanoparticles that are

© The Author(s) 2015
A. Bandyopadhyay et al., *Nanoparticles in Lung Cancer Therapy - Recent Trends*,
SpringerBriefs in Molecular Science, DOI 10.1007/978-81-322-2175-3_5

designed for transfection of genes to tumor cells are positively charged liposomes, solid lipid nanoparticles, and some polymeric nanoparticles. However, positively charged nanoparticles may also induce acute hemolysis and make the patient anemic. PEGlyation of nanoparticles may effectively reduce their interaction with blood proteins and other opsonins, thereby increasing the half-life of the nanoparticles in human circulating system. In fact, most of the nanoparticles intended for biological applications are PEGlyated as a safety measure. Treatment of lung cancer may be further enhanced by use of multifunctional hybrid nanoparticles. If researchers design an anticancer tool at nanometer scale which is functionalized with many groups intended for conjugation with imaging probes and two or more chemotherapeutic drugs simultaneously, then the patients' diseases may be detected very fast and cured instantaneously. However, again multifunctionalization of nanoparticles may further complex the situation in understanding of their behavior in human body. So far Aurimune (CYT-6091) is the only multifunctional nanoparticle that has been tried clinically [1]. Aurimune consists of tumor necrosis factor (TNF) bound to GNPs for simultaneous imaging and therapy. Obviously, multifunctional nanoparticles are highly demanding in the treatment of lung cancer where a patient hardly gets any time for treatment as lung cancer spread very fast to other vital organs. It is recommended that if nanomedicines pertaining to lung cancer therapy are manufactured as personalized medicines (for therapy of individual patients) may improve the therapeutic efficacy of nanoparticles. Often for the treatment of life-threatening diseases, personalized medicines are developed on the basis of his/her genotype. Individual genetic structure often affects any standard treatment and patients do not respond to them. In the development of personalized medicines, physiological characteristics and genetic pattern of individual patients are studied. Hence, this may give a clear idea how a particular therapeutic molecule may behave in his/her body. Definitely, researchers will be able to develop nanoparticles with appropriate functionalization at their surfaces to enhance therapeutic index of nanoparticle-based drug/gene/imaging probe delivery vehicles without causing any damage to human internal system. Even inorganic nanoparticles with very high intrinsic therapeutic value may be safely used in human body.

There are few ethical issues, which must be considered as powerful barriers in the commercialization of nanomedicines. In many places, standards in the manufacturing processes of nanoparticles are not met. Commercially available nanomedicines are often tested in a crude way for getting the license. Unethical business is often practiced by many greedy people while circulating the life-supported medicines to various stores. The problem is more pronounced in developing countries. Hence, patients either has to rely on imported medicines or they have to travel abroad for receiving proper treatments. This brings the question of affording capacity of the patients and their family. Often many patients die due to lack of treatments or wrong treatments. Many a time, doctors recommend nanomedicines to patients without any knowledge of their composition or how they may work in the patients' bodies. Patients and their families also believe more in conventional lung cancer treatment methodologies (chemotherapy, radiotherapy and photodynamic therapy, or combination of them), which are time consuming and fail to prevent cases of recurrence. They do not readily welcome nanomedicines, especially if they are the first to be tried (as most of the formulations are very new). Hence, the mortality rate due to cancer, especially of lung cancer, still remains very high throughout the world. There is an immediate need for raising concerns of the public on the fatality of the disease of lung cancer and how new nanotechnology-based targeted therapies may prove to be suitable for its diagnosis and treatment at a very early stage. There must be proper institutes, which will test whether various nanomedicinal formulations meet the required standards or not and then provide the license for clinical trials. Medical stores supplying life-supporting medicines must also be licensed and periodically monitored by the government to prevent any unethical business. Last but not the least, more and more researches have to be done for development of better nanomedicines (meticulously passing every test at cellular level), especially for lung cancer which may be tried clinically at a large scale.

Reference

1. Wang R, Billone PS, Mullett WM (2013) Nanomedicine in action: an overview of cancer nanomedicine on the market and in clinical trials. J Nanomaterials 2013:12